SOJOURNER

SOJOURNER

AN INSIDER'S VIEW OF THE MARS PATHFINDER MISSION

Andrew Mishkin

BERKLEY BOOKS, NEW YORK

The Berkley Publishing Group
Published by the Penguin Group
Penguin Group (USA) Inc.
375 Hudson Street, New York, New York 10014, USA
Penguin Group (Canada), 10 Alcorn Avenue, Toronto, Ontario M4V 3B2, Canada
(a division of Pearson Penguin Canada Inc.)
Penguin Books Ltd., 80 Strand, London WC2R 0RL, England
Penguin Group Ireland, 25 St. Stephen's Green, Dublin 2, Ireland (a division of Penguin Books Ltd.)
Penguin Group (Australia), 250 Camberwell Road, Camberwell, Victoria 3124, Australia
(a division of Pearson Australia Group Pty. Ltd.)
Penguin Books India Pvt. Ltd., 11 Community Centre, Panchsheel Park, New Delhi – 110 017, India
Penguin Group (NZ), Cnr. Airborne and Rosedale Roads, Albany, Auckland 1310, New Zealand
(a division of Pearson New Zealand Ltd.)
Penguin Books (South Africa) (Pty.) Ltd., 24 Sturdee Avenue, Rosebank, Johannesburg 2196, South Africa

Penguin Books Ltd., Registered Offices: 80 Strand, London WC2R 0RL, England

PRINTING HISTORY
Berkley hardcover edition / December 2003
Berkley trade paperback edition / December 2004

Berkley trade paperback ISBN: 0-425-19839-1

The Library of Congress has catalogued the Berkley hardcover edition as follows:

Mishkin, Andrew.
    Sojourner : an insider's view of the Mars Pathfinder mission / Andrew Mishkin.—1st ed.
        p.   cm.
    Includes index.
    ISBN 0-425-19199-0
    1. Mars Pathfinder Project (U.S.).   2. Sojourner (Spacecraft).   3. Space flight to Mars.
4. Mars (Planet)—Exploration.   I. Title.

TL789.8.U6P386   2003
629.43'543—dc22
2003061754

PRINTED IN THE UNITED STATES OF AMERICA

10   9   8   7   6   5   4   3   2   1

For Hank Moore,
who loved Sojourner most of all

CONTENTS

THE SOLAR SYSTEM AND MARS

Beyond the Moon, Earth's closest planetary neighbors are Venus and Mars. With our current technological capability, robotic spacecraft can reach either of these worlds in only months of travel. All four of the inner planets—Mercury, Venus, Earth, and Mars in order of increasing distance from the sun—are solid, rocky worlds. Traveling outward from Mars, you would next encounter the asteroid belt, remnants of once larger bodies, now consisting of thousands of pieces ranging from half a mile up to hundreds of miles across. This field of debris marks the transition to the outer planets—Jupiter, Saturn, Uranus, Neptune, and Pluto. These worlds are years away at the speeds our spacecraft can achieve. Except for Pluto, the outer planets are all gas giants, huge liquid and gaseous worlds with no discernible solid surfaces. Most of the planets of the solar system have one or many moons orbiting them; some of these moons rival planets in size, while others are mere chunks of rock and ice.

The planets are varied and distinct. Mercury is small, airless, and cratered. Venus is the greenhouse planet, with a crushingly dense carbon dioxide atmosphere and a surface temperature hot enough to melt lead. The largest planet in the solar system, Jupiter, could swallow thirteen hundred Earths. Rings thousands of miles across but only 450 feet thick girdle Saturn. Uranus is a gas giant planet turned on its side, with one of its poles sometimes pointing nearly directly at the sun. Blue-banded Neptune is nearly 3 billion miles from the sun, thirty-one thousand miles in diameter, with storms in its atmosphere as big as planet Earth. Most remote of the known planets is Pluto, solid and icy and smaller than Earth's Moon, so far away that, together with its own moon Charon, Pluto goes around the sun only once in 248 Earth years.

Far beyond the planets (months away even at the speed of light), in the deep darkness where they cannot be detected, other objects orbit the sun: perhaps billions of icy comets, composed of material unchanged since the formation of the solar system. Only rarely does a comet venture into the realm of the inner planets, where the heat of the sun vaporizes its ices into a visible tail millions of miles in extent.

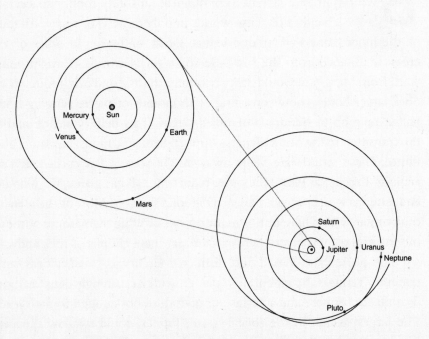

The orbits and relative positions of the planets on July 4, 1997.

Our Earth orbits the sun at a range of about 93 million miles. It is unique among the planets in having an atmosphere of mostly nitrogen and oxygen, and oceans of water covering 70 percent of its surface. Earth's Moon has no atmosphere, always shows the same face toward Earth, and is cratered with the evidence of meteor bombardment from billions of years ago.

Half again as far from the sun as Earth orbits the fourth planet—Mars. This reddish world is smaller than Earth, with only a tenth the mass, and with its entire surface about the same area as Earth's continents. Two misshapen moons—Phobos and Deimos—circle the planet. Despite its diminutive size, Mars is the home of Olympus Mons, the largest identified volcano in the solar system—rising fifteen miles into the sky—and of Valles Marineris, the grandest canyon, as long as the continental United States is wide. Much of the planet is covered with impact craters and volcanoes. There are also channels that appear to have been created by flowing water and catastrophic floods. Yet there is no liquid water visible on the surface today. Where did the water go?

The air of Mars is composed almost entirely of carbon dioxide. But this atmosphere is very thin, only about one two-hundredth the sea-level pressure of Earth's air. Although the atmosphere is tenuous, winds of up to 180 miles per hour sometimes result in regional and even planet-wide dust storms that obscure surface features for months at a time. Martian gravity is weaker than Earth's, such that a person weighing a hundred pounds on Earth would weigh only thirty-eight pounds on Mars.

Mars has seasons like Earth, and as the seasons change, its polar ice caps (of carbon dioxide and water ice) grow and ebb. The mean Martian day is just a touch longer than Earth's, at twenty-four hours and thirty-nine minutes. During the day, the temperature can rise to as high as 60°F, but even in the more temperate equatorial regions may fall at night to –90°F, with temperature swings of up to 110°F in a single day. Mars's greater distance from the sun causes it to orbit more slowly, giving it a year that is 687 Earth days long. As Earth and Mars travel in their courses, they may pass as close to each other as 34 million miles; but when their orbits take them to opposite sides of the sun, the distance can be as great as 230 million miles.

Some have postulated that Mars was once much more like Earth, but due to its smaller size, aged more quickly to become the seemingly dead planet of today. But no one knows for sure that life might not have once thrived on the planet, or that vestiges of such life might not yet remain.

Mankind has turned its attention to the exploration of Mars, for— beyond the Moon—it is the most hospitable and reachable destination in the frontier of space . . .

PREFACE

The story of Sojourner is a broad and complex one. The rover itself was small, but it represented many people who invested their lives for a number of years to make it work. As with any grand undertaking, there is no single reality shared by all involved. The story changes with the storyteller.

I have reconstructed the history of Sojourner with the aid of notes, conversations, interviews, emails, official documents, and personal memories. In some cases I have attempted to re-create conversations from my own recollections or those of others. For brevity and style, I have presented such conversations in quotes. I have always had at least one party to the original conversation review the text to establish that it represents the content of the situation accurately, when exact words are no longer available. The responsibility for any inaccuracies or misrepresentations introduced must of course be my own.

Many people contributed to the success of Sojourner. Some of them appear in the chapters that follow. Some devoted full time (and more) for all of the years of the Pathfinder mission. Some worked part time, or contributed a necessary element and moved on to the next job that needed doing. The nature of "faster, better, cheaper" is that there is less oversight: The success of the entire mission depends on each individual doing his or her job well. Despite my involvement with the development and operation of Sojourner, any story I tell must by necessity represent only a limited interpretation of the facts as colored by my personal experience. Most members of the rover team do not appear in the body of this account, and I have little personal knowledge of many of their contributions. However, the fact of their omission from this text should not be construed to diminish their importance to the mission.

A history of Sojourner must be the story of two journeys: the rover's voyage to Mars, and the shared experience of the individuals who, together, made that voyage a reality. In an attempt to ensure that the members of the rover team receive the recognition they are due, I have included below the most complete compiled list of members of which I am aware. I have not tried to encompass the entire Pathfinder mission team, but only the individuals, institutions, and companies directly participating in the rover effort. If, despite my best efforts, errors remain, I offer my apologies. To the rover and Pathfinder teams I offer my thanks for the privilege of being a part of a once-in-a-lifetime adventure. We did it!

THE ROVER TEAM

Jet Propulsion Laboratory
George A. Alahuzos
Ghanim Al-Jumaily
Teresa D. Alonso
Sami W. Asmar
Lawrence W. Avril
Ronald S. Banes
Edmund C. Baroth
Sheryl L. Bergstrom
Donald B. Bickler
Russell D. Billing
David J. Boatman
Stephen R. Bolin
Gary S. Bolotin
Thomas J. Borden
David F. Braun
Kyle D. Brown
Robert E. Brown
Dale R. Burger
John M. Cardone
Brian K. Cooper
Cosme M. Chavez
Joy A. Crisp
Evan D. Davies
Bob C. Debusk
Henry Delgado
Fotios Deligiannis
Tolis Deslis
William C. Dias
Patrick L. Dillon
Johnny Duong
Howard J. Eisen
Khanara Ellers
Brenda Fieri
Harvey A. Frank
Jack A. Frazier
Bert H. Fujiwara

Robert D. Galletly
Jeffrey A. Gensler
James A. Gittens
Willis W. Han
Christopher Hartsough
Gregory S. Hickey
Denise A. Hollert
Sean Howard
Tien Hua
Kenneth A. Jewett
Kenneth R. Johnson
Vaughn J. Justice
Faramarz N. Keyvanfar
Eug-Yun Kwack
Sharon L. Laubach
Geoffrey A. Laugen
William E. Layman
Todd E. Litwin
James W. Lloyd
Gena Lofton
Justin N. Maki
Ramachandra Manvi
William H. Mateer II
Jacob R. Matijevic
Larry H. Matthies
Thomas M. McCarthy
Robert McMillan
Donald P. McQuarie
Robert J. Menke
Marian G. Meridieth
Andrew H. Mishkin
David S. Mittman
Gordon R. Mon
Robert G. Moncada
Henry J. Moore
Yvonne Morales
Ronald A. Morgan

Jack C. Morrison
Robert L. Mueller
Fred F. Nabor
Michael A. Newell
Tam T. Nguyen
Don E. Noon
Timothy R. Ohm
Argelio Olivera
Thomas O'Toole, Jr.
Mervin K. Parker
Michael D. Parks
Douglas C. Perry
Mark Phillips
Betty L. Preece
Thomas A. Rebold
Evelyn Reed
Andrew D. Rose
Robert E. Scott
Terri A. Scribner
Leonard A. Sebring
Dee R. Sedgwick
Cesar A. Sepulveda
Donna L. Shirley
Allen R. Sirota
Hugh Smith
Kathy Sovereign
Beverly St. Ange
Christopher B. Stell
Paul M. Stella
Philip P. Stevens
Henry W. Stone
Scot L. Stride
Lee F. Sword
Hung Ta
Jan A. Tarsala
Robert F. Thomas
Arthur D. Thompson

Joseph F. Toczylowski
Peter Tsou
Lin M. van Nieuwstadt
Matthew T. Wallace
Yolanda Walters
Wilson "Bud" F. Watkins
Richard V. Welch
George H. Wells, Jr.
Liang-Chi Wen
Brian H. Wilcox
Larry Wild
Paul B. Willis
Rosalinda Wilson

Lewis Research Center
Dale C. Ferguson
Phillip P. Jenkins
Joseph C. Kolecki
Geoffrey A. Landis
Robert J. Makovec
Lawrence G. Oberle
Steven M. Stevenson
David Wilt

APXS Instrument Team
Richard S. Blomquist
Frank DiDonna

Pasquale DiDonna
Thanasis E. Economou
Holly A. Kubo
Maury Perkins
Michael J. Phillips
Rudi Rieder

Contractors
Astro Aerospace
    Corporation
Beckman Instruments
Canon Connectors
Castrol
Data Radio Inc.
Department of Energy
Environmental Test
    Laboratory
Eastman Kodak/
    Microelectronics
    Technology Division
Falcon Designs
Garwood Laboratories
Globe Motors
Hewlett-Packard Van
    Nuys
Interelectric A.G.
Litronic Industries

Lucas Schaevitz
Maxon Precision Motors
Minco
Motorola Derivative
    Technologies Division
Motorola Radius
    Division
National Semiconductor
Pacific Scientific
Pico Electronics
Pioneer Circuits, Inc.
Power Trends
Rosemount Aerospace
Saft America
Spectra Diode
    Laboratory
Systron Donner
Tecstar/Applied Solar
    Division
Telogy
Wyle Laboratories

SOJOURNER

PROLOGUE

April 1985

"I can't claim to be an expert on Artificial Intelligence."

The light in the office was dim. I was splitting my attention between the two interviewers. The dark-haired man to my left smiled. "We'll learn about it together," Neville Marzwell said with an indeterminate accent. He seemed kind, yet somehow mysterious.

Ed Kan, sitting directly across the table from me, was ready to finish things up. He had a project that should have started some time before, but hadn't. Marzwell had people working his own project, but needed more. "We'd like you to start Monday."

It was my second interview at the Jet Propulsion Laboratory.

I had been working at my own company for about three years. My partners were mostly brilliant but temperamental engineers. We had been designing hardware, doing software consulting, and writing venture capital proposals that never got funded. When I decided I wanted a regular paycheck, I began sending out résumés and wondering who would hire a systems engineer a few years out of school whose only experience since was as a struggling entrepreneur.

One of my first interviews was at JPL in Pasadena. I got a call from an old friend I'd worked with when we were both graduate students at UCLA a few years before. He was moving to the east coast, and his position in the JPL Advanced Teleoperation group was opening up. This group worked on Space Shuttle flight experiments, designed robotic manipulators to be operated remotely either by astronauts in space or people on the ground, and researched how to control these devices effectively from hundreds or thousands of miles away. My friend was recommending me for his position, and would I like to come in and meet with his supervisor? I went in, met Ed Kan and others, had lunch in the JPL cafeteria with his robotics group.

The interview had seemed to go well. But JPL didn't call. I sent out more résumés. After a few weeks I made the "follow-up" call. "We decided not to fill the position," Kan told me.

Two months later, copies of my résumés were completing their meanderings through the labyrinths of several companies' personnel offices. The number of callbacks increased. Amid this, Ed Kan called me. "We'd like to do a second interview." How soon? "Tomorrow."

Ten days later I started work. I could have waited for other offers. I almost certainly could have wrangled a higher salary somewhere else. But I'd always wanted to be in the Space business. And for that, JPL was the place to be.

July 5, 1997

For the first time, the Sojourner rover's six cleated wheels had made tracks in the butterscotch-hued Martian soil of Ares Vallis. Nearby sat the Pathfinder lander, Sojourner's home and protector during the seven-month voyage to Mars. Inches from the rover's wheels was the end of the ramp down which it had just driven. It was mid-afternoon: a dimmer, smaller sun was halfway down the sky.

On Earth, the rover team was ecstatic. Images of the rover had just arrived from far away. The Mission Control area was wild with cheers and applause. ". . . six wheels on soil!" announced the Rover Coordinator.

Sojourner was fulfilling its promise. We now had a fully functional, healthy rover on the surface of an alien world. The rover team was ready to drive the machine we had built and trained to operate. As a member of that team, I was both awed and humbled by what we had done, and what we were about to do.

It was time to go exploring.

PART 1

LAYING THE GROUNDWORK

ONE

DOING WHAT'S NEVER BEEN DONE

Out beyond the orbit of Neptune it is cold. Very cold. The sun is a bright star. Even out this far, there are charged particles of matter streaming from the sun. This tenuous solar wind becomes ever weaker the farther one travels into deep space. Eventually, the flow meets and is halted by particles from outside the solar system—the interstellar wind. Astronomers call the region where this happens the "heliopause." As distant as this realm is from human experience, through it fly artifacts made by human hands. Voyagers 1 and 2 are traveling farther and farther from home, searching for the heliopause, each week sending back radio reports of their latest instrument observations. Home is the Jet Propulsion Laboratory.

One of the key objectives of JPL is simply "the robotic exploration of space." Other NASA centers launch human beings into space to orbit the Earth, and have sent men to the surface of the Moon. JPL sends scouts into the solar system where humans cannot yet go.

If you drive a few miles west on the 210 Freeway out of Pasadena, California, you will soon notice a complex of structures off to the right. The JPL facility is a sprawl of buildings located at the base of the foothills of the San Gabriel Mountains, surrounded only by residential neighborhoods and undeveloped hillsides. It is here that missions to the planets are designed, implemented, managed, and controlled. And, perhaps most renowned, here is where engineers troubleshoot spacecraft that are already en route to unimaginably distant destinations, and invent workarounds to keep missions alive when components fail and the nearest repairman is hundreds of millions of miles away.

The atmosphere at JPL appears relaxed. The facility is campuslike, with over a hundred buildings spread out over a large area, and a mall by the main gate with greenery and fountains. There are few tall buildings. The highest, at nine stories, is the main administration building on the north end of the mall.

JPL is about Space. There are a few streets that link the site, and they are named for spacecraft that came from here: Surveyor Road, Explorer Road. The embroidered shoulder patches on the security guards' uniforms boast Saturn and its rings. The hall of the first floor of the Space Flight Operations Center has been transformed into an art gallery of sorts. A typical company might exhibit pictures of its products, but JPL displays images of planets and moons photographed by its spacecraft, the unprecedented detail in each one representing a new scientific discovery. On these walls a visitor can see radar maps of the surface of cloud-enshrouded Venus, or volcanoes erupting into space on Jupiter's moon Io.

Despite its current reputation, JPL was not established to explore space, and the notion of doing so was, at the time of its birth, considered pure fantasy.

※

In 1926, the California Institute of Technology received a $300,000 grant to establish an aeronautics laboratory and graduate school. Over the next two years the Guggenheim Aeronautical Laboratory of the California Institute of Technology—known as GALCIT—was built, and in 1930 Theodore von Karman became its director. It was out of the activities of von Karman and GALCIT that JPL would arise.

One of von Karman's graduate students, Frank J. Malina, proposed to examine the problems of rocket propulsion for his dissertation. The general academic attitude of the time was that rocketry had no merit and belonged to the realm of pseudoscience, yet von Karman okayed the research. Together with two enthusiasts who "wanted to fire rockets," Malina assembled a rocket motor within a year. They were ready for their first test on Halloween 1936, with the rocket engine mounted on a test stand and the nozzle pointed up at the sky. The researchers picked a location in the dry riverbed of the Arroyo Seco, some six miles northwest of the Caltech campus. The riverbed was largely devoid of flammable vegetation and seemed a good spot to set off a rocket engine. The group huddled behind piled sandbags and ignited the motor—which shut down instantly. The first test had failed. So did the next several attempts. But by January, they succeeded in operating their rocket engine for nearly forty-five seconds.

A few other Caltech graduate students became interested. The small GALCIT rocket group continued theoretical analysis, construction and test of simple rocket engines. The group funded its efforts almost entirely out of the pockets of its members.

As World War II loomed, government funding flowed into GALCIT. The rocket group moved off the Caltech campus, relocating to the site of the 1936 rocket tests in the Arroyo Seco, on land leased from Pasadena. (Fifty years later, robotic rovers would be tested in the same spot.) Through a combination of theory, experimentation, failures, and inspiration, the growing team developed the first practical rocket engines in the United States. They were called JATOs, for Jet Assisted Take-Off. During the war, JATOs helped airplanes get from the decks of aircraft carriers safely into the air.

*

On October 4, 1957, the Union of Soviet Socialist Republics placed into orbit the Earth's first artificial satellite, named Sputnik. JPL partnered with the Army Ballistic Missile Agency to respond within three months with the first successful launch of a U.S. satellite—Explorer 1. Soon after, control of JPL transferred from the U.S. Army to the newly created National Aeronautics and Space Administration. JPL managers expected to continue development of new rockets and space propulsion systems.

But by 1960, JPL's focus had shifted almost solely to the spacecraft that would ride those rockets to the Moon and the planets. In the early years of the decade, JPL sent space probes to hard-land on the Moon and through interplanetary space to fly past Venus.

Other JPL missions followed in the late sixties and beyond. Surveyor, built for JPL by Hughes Aircraft Company, soft-landed on the lunar surface in advance of the Apollo human landings. The Mariner missions flew past Venus and Mars. In 1973, Mariner 10 was the first spacecraft to closely explore Mercury, coming closer to the sun than any previous probe.

Viking was the first mission to reach the surface of Mars, sending two orbiters and two landers to the planet. It was also the first planetary exploration mission with a billion-dollar price tag. JPL found itself in a rare support role, with another NASA facility, the Langley Research Center in Virginia, having ultimate responsibility for Viking's success. Langley came to JPL for the Mars orbiters, and to the Martin Marietta Aerospace Company for the landers.

JPL proposed the Grand Tour of the outer planets. Due to a fortuitous alignment of Jupiter, Saturn, Uranus, and Neptune that would take place in the late 1970s, but not again for hundreds of years, it would be possible to send a single spacecraft skipping from one planet to the next, collecting data from them all. By taking advantage of the gravity slingshot effect possible at each planetary encounter, the spacecraft could reach the next planet in line much sooner than by a dedicated direct trajectory. The budget for the Grand Tour was not forthcoming. JPL scaled it down to Voyager, and did it anyway. The Voyagers would prove so reliable that the spacecraft would still be functioning twenty-five years later.

Deep space missions grew ever more ambitious. Galileo—another billion-dollar-class mission—was not to just pass near Jupiter, but to go into orbit, exploring the entire Jovian system of moons. As if this weren't enough, Galileo would release a probe that would parachute into Jupiter's atmosphere, dropping deeper and deeper until it was crushed by the increasing pressure and finally vaporized by the planet's intense heat.

The other JPL space probes built in the late eighties and early nineties approached Galileo in complexity, cost, or ambition. Magellan was to orbit Venus, mapping its surface. Since cameras could not see through the dense Venusian clouds, Magellan would use radar imagery to map Venus to a resolution even better than yet achieved for Earth by satellites orbiting our home planet. Mars Observer was planned to circle the fourth planet, carrying many science instruments, taking pictures that in some cases would distinguish features less than six feet across. And the Cassini mission would do for the Saturnian system what Galileo would do for the Jovian.

Even as Cassini was being approved by Congress to begin development, the political and economic climate was changing. The United States budget was tightening. A billion dollars for a planetary mission was just too much. Weren't there better, more important ways to spend the money?

During the first twenty-five years of the space program, the American public had perceived NASA as perhaps the only truly competent government agency. On January 28, 1986, this view faltered. The Space Shuttle Challenger exploded seventy-three seconds after launch, killing all members of its crew. The shuttle fleet was grounded for over two years. But NASA's reputation was grounded for years to come. NASA's vaunted "safety first" policy was made a lie. Other space failures followed, including a string of unmanned satellite launches that either blew up or failed to achieve proper orbits. Some of these had no connection with NASA, but this distinction seemed to go unnoticed in the public eye. When the Space Shuttles finally began flying again, the Hubble Space Telescope was deployed into orbit, with high hopes that it would enable astronomers to see farther into the depths of the Universe than ever before in history. Yet only a few weeks after launch, the Hubble was found to be "nearsighted" due to a manufacturing error.

JPL was not immune to high-profile problems. The Galileo spacecraft, intended for launch from the Space Shuttle in 1986, was delayed for years in the aftermath of Challenger. When it was finally launched, its main "high gain" antenna, designed to open like an umbrella, became stuck only partially deployed. The antenna was useless. Without it, only a small fraction of the planned high-resolution images of Jupiter and its moons would ever be sent back to Earth. A likely cause of the problem was identified: lack of lubrication of the antenna deployment assembly. The engineers who designed the antenna had not anticipated the delays in the launch of Galileo. In the original schedule, Galileo would have launched and the antenna would have been opened within a year of its construction, and the initial lubrication would have sufficed. Instead, over three years had elapsed before launch, and the lubricant had dissipated.

Mars Observer was to be the first return to the Red Planet in fifteen years. Both the launch and the cruise to Mars proceeded without incident. But just hours before the spacecraft was to fire its thrusters to put itself into orbit around Mars, it stopped communicating with the Earth. Ground controllers at JPL attempted to reestablish a link for months, but there was never an answer. JPL had lost a spacecraft for the first time in twenty years.

In the middle of this gloom, the Magellan mission was a success, mapping over 95 percent of the surface of Venus, outliving its expected mission duration.

NASA's budget had been shrinking for years, ever since the late Apollo days. Congress's support of space exploration was at best halfhearted. The NASA Administrator responded with a new mantra for future space efforts: they must be "faster, better, cheaper." NASA must do more with less. These seemed at first to be empty buzzwords. Instead, they signaled recognition of what had to be done differently, perceived differently, if NASA was going to survive.

Now JPL would have to do more with less as well. The organization could build spacecraft to perform unprecedented technological feats. Each mission into the solar system proved it. But simple technical excellence would no longer be enough. The winds had shifted. What would this mean to the future of JPL? More was riding on its actions than ever. The "faster, better, cheaper" environment now challenged JPL to again do what had never been done before.

TWO

"ALMOST AS GOOD AS A BOGIE"

Don Bickler was looking for something new.

After the oil embargo of '73 there had been a lot of interest in alternative energy sources. That interest had translated into money. Bickler had been hired at JPL in 1975 to work on ways to bring down the costs of solar cells. The program was big—over a hundred million dollars per year. As part of the project, JPL sought to improve the process of manufacturing solar cells, including refining the silicon, slicing it into wafers, making cells from the wafers, and finally constructing the panels from quantities of individual cells. Bickler's specialty was the production of the cells from the sliced wafers. By 1985 JPL had done it. In 1975, the cost per panel was about twenty dollars per watt. Ten years later, adjusted for inflation, it was down to just a dollar a watt. And the money for the program was starting to dry up.

There was other work for Bickler at JPL. But he kept digging around for something to pique his interest. About this time, his kid bought a Jeep. Just like he did with any mechanical system he came across, Bickler studied the Jeep and the available after-market modifications to its suspension to see what made it work, and how well it performed. He was disappointed. "I could do better than this."

It might be intriguing to attack the problem of designing high-mobility vehicles intended to go off-road. Bickler began to think that maybe it was time for the world to get interested in extraterrestrial vehicles. Extraterrestrial, by definition, meant way off-road.

Bickler knew that there was ongoing research at JPL into robotic vehicles. Carl Ruoff, who had been supervisor of the Robotics group, had managed to capture some of JPL's scarce internal research-and-development dollars and direct them toward developing some of the fundamental technologies required to make "planetary rovers" possible. Bickler had once given Ruoff funding to buy a robot arm to experiment with robotic assembly of solar cells. Maybe Ruoff wouldn't mind a little help with rovers.

The trouble was that Bickler was in the Mechanical Systems division, and Ruoff worked in the Electronics and Control division. Robotics was a perennial bone of contention between the two JPL organizations. In an institution of six thousand people constantly developing new technologies and applications for those technologies, gray areas in the management hierarchy were a fact of life. The electronics and control engineers thought robotics was their domain, because making a robot do the right thing required that software and control algorithms be combined with sensors and motors into a single integrated system. The mechanical people saw the robot as a complex electromechanical system composed of motors, gears, and linkages that needed to be designed elegantly to work together: "Those control guys always want to write software to solve a problem that would be trivial if they had just designed the hardware right in the first place . . ." And robotics was one of the sexy technologies. Management in both divisions wanted to dominate robotics activities at JPL.

Bickler didn't care about any of that. He had found an unresolved engineering research topic to delve into. He just wanted the opportunity to do it.

Amid the technological saber rattling, Bickler had a meeting with Ruoff. Ruoff was an even-tempered man with a keen sense of the absurdity of the foibles of large organizations. (He would sometimes pat half-inch-thick copies of viewgraph presentations he had been forced to either sit through or present himself, and then comment wryly: "Ah, viewgraphs. Our most important product!") When Bickler told him what he wanted to do, Ruoff said, "Well, Don, in spite of your management, I'll cooperate with you." Ruoff told him to go see Brian Wilcox, who was co-ordinating the planetary rover research effort.

So Bickler walked across the Laboratory to Wilcox's office. The building was about as far from the main entrance of JPL as you could be and still be on-site. That also put it far from the administrative offices, thus reducing the number of visits from management. This was where the Robotics group did its tasks in skunkworks-like isolation.

Wilcox had a thick black beard that covered most of his face. He had recently been promoted to supervisor of the Robotics group even though he was far more interested in exploring creative ways to make things work than in moving further up the management chain.

The rover researchers could not afford to build a new rover for their experiments. Instead, they had refurbished a rover that was already available: the Surveyor Lunar Roving Vehicle, or SLRV. This six-wheeled rover had been constructed in the early 1960s and was designed to fit onboard one of the unmanned Surveyor spacecraft that were being soft-landed on the Moon in advance of the Apollo manned landings. The Surveyors had proven that safe landings were possible, and that the first lunar astronauts would not disappear forever beneath an ocean of dust. The Surveyor designers had also hoped to send a lunar rover, operated from the Earth, to scout the surface. Two prototype SLRVs were built, but none were ever sent to the Moon. With a few thousand dollars, Wilcox's team had pulled one SLRV out of mothballs, installed new batteries, cameras, and tires, and made the vehicle operational again. They also painted over the original white color with a light blue. From that point on, the vehicle was known simply as the "Blue Rover."

What Bickler wanted to do was to optimize the mobility performance of a planetary rover. Could he come up with a design that would be better than the Blue Rover at driving over rocks and crossing crevasses? How could he minimize the chances that a rover would get stuck on a rock or sink into sand? Was six the right number of wheels? He did not want to compete with the current research activity. What he asked Brian Wilcox was "Is there anything I can do without mowing your lawn?"

Wilcox responded that Bickler was welcome to investigate rover mobility characteristics. Wilcox had much more important things to worry about. His team had never signed up to improve the Blue Rover. They were deep into the issue of how to control a rover that was so far away— say, on Mars, for instance—that there was no way to drive it directly. Suppose the rover had a TV camera mounted on it, sending back pictures of the terrain in front of the rover. Traveling at the speed of light, those TV images could take up to twenty minutes to reach Earth. Commands from a driver on Earth would take another twenty minutes to get back to the rover where they were needed. So any instructions to the rover, such as "Stop before you go over that cliff!" would arrive forty minutes too late. Time delay, even at the speed of light, could be a killer. How did you deal with it? Somehow, the rover had to be made to go to the right place, without a human being immediately available to tell it what to do.

No matter how mobile the vehicle might be, there could still be rocks big enough to get stuck on, and crevasses wide enough to fall into. On the other hand, the more capable the vehicle's mobility, the more types of terrain it could handle safely, and the more interesting places it could go. And Wilcox was happy to have Bickler work with gears and motors and come up with the best rover design he could.

Ruoff had already given Bickler a copy of a book by M. G. Bekker, the inventor of the Blue Rover. Wilcox also had a video of the original mobility tests of the SLRV from the mid-sixties, after it had been delivered to JPL, which he let Bickler borrow. Bickler went away to study and learn.

✳

Bickler liked to design and analyze. But he did not like to read books from beginning to end. He flipped back and forth through the Bekker book. What made the Blue Rover so special?

It was a six-wheeled vehicle, with all wheels independently electrically driven, and divided into three bodies. The front, middle, and rear compartments were each supported by two wheels. The three compartments were linked by a spring-steel member. For steering, the front and rear compartments could rotate on their attach points at the center between the two wheels. To make a right turn, the front body would rotate clockwise, the rear counterclockwise; then the entire vehicle would drive forward in an arcing right turn. When the rover had traveled far enough along its arc to be facing in the desired direction, it would stop, straighten the front and rear compartments, and drive forward again in a straight line. The six-wheel design meant that even when one compartment with its two wheels was negotiating a hazard, the vehicle still had four powered wheels firmly on the ground providing traction and stability. If the front wheels were handling an obstacle, the rear four were providing the "push." By the time the obstacle got to the middle wheels, the front and rear were the stabilizing influence, and so on.

It was magnificent! Nothing could touch it. The Blue Rover could drive over rocks one and one-half times as high as its wheels. That would be like being able to drive your car over your dining room table! Because the three bodies of the rover were connected to each other by a spring, each body could twist relative to the others. As a result, even in rough terrain, all six wheels usually stayed on the ground.

One day while looking through Bekker's book, Bickler came across a passage that intrigued him. It said that a sprung suspension, like the Blue Rover's, could be "almost as good as a bogie." Now, what did that mean?

A bogie suspension has no springs. Instead it relies on rigid linkages that can pivot relative to each other. Bogies were used on locomotives, some trucks, and in military tanks.

So there were limitations to the Blue Rover design. If the front wheels began to climb a rock, the spring-steel linking the front body to the rest of the vehicle would be forced to bend. Being a spring, it tended to resist bending, and that meant that the forward wheels pushed down harder onto the rock than they would if applying just their own weight. Bickler mulled this: That was the opposite of what you really wanted to do. If you could take more weight off the wheels going over the obstacle, it would be easier to get them past it. Putting more weight on the wheels climbing the rock made negotiating the rock just that much more difficult. Another problem was that when more weight went to the front wheels, less went to the middle and rear wheels, giving them less traction. And traction on the other wheels was just what was needed to help get the front wheels over the hump.

Bogies might be a way to allow the various wheels to conform to the terrain, but without shifting weight to the wrong wheels, like the Blue

Rover did. That must be what Bekker was getting at. Maybe that one sentence in Bekker's book meant that there was something Bickler could bring to Bekker's party. When rovers finally did get to other planets, they would need all the mobility performance that could possibly be squeezed out of a design.

Bickler grinned. Bogies!

※

Don Bickler grew up in Chicago, and went to college there to study mechanical engineering. While in college, he became a co-op employee at Stewart-Warner, a manufacturer of automotive equipment. He was immediately put out on the factory floor. Every week he would be moved from one factory job to the next so he could be familiarized with each one. He worked with the sheet metal benders, then the welders, then the mill operators, and so on.

Early on, the foreman took Bickler aside and told him, "You've got to take your turn at the trash heap." The trash heap was a pile of leftover stock from the machinists. The pieces, each about ten feet long, almost completely covered a workbench. They spilled over on the floor, under the workbench, and encroached on nearby work areas. The stock could not be used, but it was too long to be loaded into boxes and shipped to the scrap yard. "What you've got to do," the foreman told Bickler, "is take this hacksaw and saw each big piece into smaller pieces, then put the pieces in the box so we can send it out." It would probably take ten minutes to saw all the way through one piece. That should keep him busy for a while. Bickler went to work. The first thing he noticed was that the stock's cross section was not round, but hexagonal. That meant he could clamp it in the vise and it wouldn't twist out. That made it a lot easier to work with. One of Bickler's recent engineering classes had been "Strength of Materials." He thought about whether anything he had just learned would apply here. Well, since this was machining stock, it had to be ductile to make it easier to machine. If it was ductile, then . . . Bickler clamped one piece in the vise, so that about eighteen inches worth stuck out one side. He grabbed the end and pulled it toward him, bending it around. Then he took the hacksaw, but instead of trying to saw all the way through, he just scored the steel near the point where it came out of the vise, twisted, and the piece in his hand came away clean, breaking off at the score point. He repeated the process on the next piece of stock. He began to get into a rhythm: Clamp. Bend. Score. Twist. Clamp. Bend. Score. Twist. He did this for a while. He seemed to be making a dent in the trash pile. After a while longer, he had converted most of the pile into shippable lengths. About this time, he noticed that the shop had gotten quiet. He stopped and turned around. There were fifteen machinists standing in a line watching him. One of them shook his head. "We've had that trash pile for twenty years," he said. "Sure, we cut it down a bit when we run out of room, but then it overflows again. You just took care of it all at once."

Bickler could only muse, "I guess engineers have their uses." If you applied what you learned from books and professors, you could often find solutions that eluded others. Building on the knowledge and codified experience of engineers who had gone before him could be a magical thing.

By the time Bickler had rotated his way through all of the positions, he had earned the goodwill of the machinists and technicians on the factory floor. So when Bickler got his first sailboat . . . all of its fittings were custom made in pure bronze and chrome-plated by the expert machinists of Stewart-Warner.

*

As Bickler hunkered down to figure out how a bogie suspension might beat the Blue Rover in off-road performance, those days in Chicago were thirty years behind him. His hair was turning silver, though he still had all of it. He had a wry smile that telegraphed *What are you trying to pull?* to any unfortunate engineer who tried to sell an idea based more on wishful thinking than true understanding. He still had the young engineer's straightforward attitude that caused him to argue passionately for the best engineering solutions based solely on technical merit, having never acquired the political sensitivity with which most engineers were afflicted after enough years in business. And he still possessed the same instinctive drive to uncover the better solution to the problem in front of him, and to build on the foundation provided by those who had examined similar problems in the past.

Bickler found himself more and more involved in the area of rover mobility design. He heard about a meeting that sounded interesting, so he attended. Ken Waldron, a noted researcher in the field of robotic vehicles, working at Ohio State University, had been funded by one of JPL's program offices to study six-wheeled vehicles and their potential performance. Waldron was visiting JPL to present his plans for review. Bickler sat in on the review to see what he might learn. When he didn't understand something, he asked questions. Soon after the review, he was asked to be the JPL contract monitor for the Ohio State contract. Maybe he did know something after all . . .

And then he got invited to another meeting. It was an organizing meeting for a workshop on "Mars Rover Sample Return" or MRSR for short, which would examine the various technologies that would make it possible to bring samples of rock and soil back from Mars. "I didn't know what I was doing in that room." Soon it became clear. The intent of the workshop was to bring together from around the country as many of the experts on rover-related technologies as possible. Brian Wilcox was already going to be leading a session on local navigation and hazard avoidance. He wanted Bickler to chair the session on rover mobility concepts. Bickler protested: He didn't even know who the vehicle design gurus were. Wilcox persisted: "Don't worry about that. I'll give you the names and numbers of all the right people. You just have to pick up the phone."

And that's what Bickler did. "So I get on the phone, and the first guy I call up, he says, 'I'm not coming unless so-and-so's coming.' So I call up the next guy, and he says he's not coming unless the first guy's there. So I'm calling back and forth, and finally get a pretty good group together. So I end up chairing this session. And that's how I got into this mobility business."

That wasn't entirely true. At just about the same time, Bickler generated his first bogie-based mobility concept. He made his first model out of plywood. The six wheels were the leftovers you got when you used a hole saw to put a hole in a door to install a doorknob. The model was simple, without motors. Just wheels and wooden linkages between them, so you could push it around on a tabletop. But even as simple as it was, you could see that it could roll over a block of wood higher than its wheels without tipping over. You could lift any wheel off the ground, to the limits of the pivots, without any other wheel moving at all; that meant it would be stable in rough terrain. Unlike the Blue Rover, the model kept equal weight on all six wheels, even when climbing over an obstacle. The only problem with the design was that you couldn't steer it. If you tried to install steering pivots to let the wheels turn, the linkages would ruin the ground clearance of the vehicle, so you really couldn't deal with the rough terrain you originally thought you could.

A year later Bickler had solved that problem, at least on paper. The new design incorporated "virtual" pivots. Using a number of four-bar linkages, the new design acted just like the old wooden model, except that all the links sat above the wheels, at a height that didn't interfere with ground clearance. Only one thin link went down to the axis of each wheel. To make each wheel steer, you just had to put a rotational joint into each of those thin links. One engineer who looked at Bickler's new design said the four-bar linkages reminded him of a "pantograph," a drafting device used to copy a drawing at any desired size by tracing it. The name stuck: People started calling the new design the "Bickler pantograph."

So far, Bickler only had his sketches of the pantograph. He was ready to build one, but he wanted this one to be motorized. He would need money from somewhere for parts. By now, mid-1988, there were more detailed studies going on to define possible future Mars rover missions. Donna Shirley, who was leading the MRSR study team, gave Bickler some funding to continue looking at rover mobility concepts, but no money to build anything. Bickler wanted to prove that the pantograph would move as well as his paper analysis said it would. But Shirley told him no: "We're not building any hardware for this study. We're only examining the 'tradespace' of options and developing a conceptual design." Bickler was stymied. He could not misdirect Shirley's funds to do something she specifically forbade; but endless evaluation of "tradespaces" was just not his idea of engineering.

One day, Shirley was presenting the results of the rover study. The audience was a review board composed of high-ranking JPL and NASA representatives. One of the slides Shirley showed was a sketch of the pantograph. A review board member stopped her and said the pantograph seemed overcomplicated to him. "What's the use of all that complexity?" he wanted to know. Bickler was sitting in the back of the room, and he dutifully chimed in with an explanation of the mobility performance of the design. The reviewer was not convinced. So Bickler asked, "Would you like to see a model of this design that you could run around on the tabletop? Then you could really see what the pantograph can do." The reviewer, along with the rest of the board, leapt at the offer.

Bickler had what he wanted. The review board outranked Shirley. He figured that he had just been given an endorsement to build a pantograph model.

With scant funding, Bickler built as much of the pantograph as he could at home in his garage workshop. Instead of complete blueprints generated by the JPL design room, he gave the JPL machine shop 8½-×-11-inch sheets of paper with drawings hand done with a felt-tip marker. And the few parts he asked the shop to produce were mostly not finished components, but merely aluminum stock cut to appropriate lengths. He bought electric motors at the local surplus store. Bickler even managed to construct fiberglass "dome" wheels in his garage. The wheels were shaped so that, on hard surfaces, the rover would ride on the rims, but when the vehicle got into soft sand, it would start to sink. More of the wheel would then come into contact with the ground, providing more surface area and helping to prevent further sinkage. It was like the difference between walking in snow in boots or in snowshoes: The person in boots might sink in up to his neck, while the guy in snowshoes could walk along nearly on the surface.

At least, that was the idea. Bickler would have to assemble the complete pantograph and give it a try. Once put together, the model was about two feet long, and had five-inch-diameter wheels. There was a motor in each wheel. The motors were ganged together, and wires led back to a battery and a switch, so the rover model could be driven forward or backward. When he first tried out the vehicle, the wheels were too slick, and would slip on hard floors or tabletops. Bickler slipped rubber bands over each wheel to give them more traction.

Bickler's analysis had indicated that the pantograph should be able to climb steps one and a half times the height of its wheels, just like the Blue Rover. So he had the carpenter shop make him a carpeted step the appropriate height. The carpeting would minimize any wheel slippage, and allow the pantograph to demonstrate just what it was capable of. When the step was ready, he went over to the shop and grabbed it. Now he'd see what the design could do!

Well, soon, anyway. Between phone calls and other interruptions, he just wasn't getting a chance to try out his new vehicle. By the time things quieted down, it was after five o'clock. He pushed the step against the wall and placed the six-wheeled model in front of it. The pantograph went up the step with ease! It could climb steps as high as the Blue Rover could. Bickler wanted to show someone what he'd accomplished. He ran out into the hall. Most everyone had already gone home. Across the hall was the division manager's office, and he was still in. So for the next hour they played with the pantograph, driving up and down the step, trying it out in various locations, steering the wheels, like two kids with a new toy on Christmas Day.

With a few more experiments Bickler was able to prove that the pantograph could cross crevasses as wide as 40 percent of the length of the rover. And the pantograph design gave it ground clearance three times higher than the Blue Rover, so it could go over much rougher terrain. All six wheels could be steered, so the pantograph could either turn like the Blue Rover, or instead turn all wheels in the same direction, and move off to the side like a crab.

※

After playing with the pantograph some more, Bickler discovered a drawback in the design: It had a problem with "bumps." If the rover came to a step, it had no trouble; but the front wheels never dropped down to their initial level. On the other hand, if the pantograph drove over a bump—like a lone rock—the front wheels ended up at the same level after going over the bump as before. The middle wheels would then have greater difficulty crossing the same bump, and with large enough bumps would not be able to get over the bumps at all. Bickler was disappointed. Why would the pantograph design be worse than the Blue Rover in this particular way? Surprise! The Blue Rover had the same problem! Bekker must have known this, but never mentioned it in his book. And if you knew what to look for, you could see the SLRV getting stuck on rocks in the videotape, then backing off and going another way. Bekker had kept a secret!

✳

The design of the high-mobility rover chassis did not end with the pantograph, or "Bicklermobile" as it was sometimes also called. The MRSR review board's objection to the pantograph had not gone away: The pantograph was complex. It had a lot of parts. Those parts added weight and risk. It cost thousands of dollars per pound just to launch a payload to Mars. If the total mass of the payload were too high, then the launch rocket would simply not be able to reach its target. So every deep space mission is pushed to keep the weight of its spacecraft low. Every subsystem is given a mass allocation, and every assembly within each subsystem is given its own fraction of that allocation. And a heavy rover would mean that fewer science instruments could be flown to Mars.

With deep space robotic missions there is no one to repair equipment failures during the mission. While clever engineers can, at some cost, design backup systems, one of the best and cheapest ways to make equipment reliable is to make it as simple and elegant as possible while still getting the job done. The multiple four-bar linkages of the pantograph just looked like a mechanism that could jam or break, begging the question "Is there a simpler way?"

The answer to that question would be called "Rocky."

THREE

OFF-ROADING WITH NO ONE
AT THE WHEEL

The white paint was ratty, mottled, and cracked. The rubber covering the spring-wire-loop wheels had rotted through, collapsing the wheels and creating the appearance of some type of mechanical beast prone in its black nest. The last time Brian Wilcox had seen the Surveyor Lunar Roving Vehicle prototype, it had been in much better condition. But that had been over twenty years before. He had been twelve years old.

In the late 1950s Brian's father, Howard Wilcox, was program manager for a classified program at the Naval Ordnance Test Station at China Lake that launched rockets from a jet fighter. Launching from a high-flying aircraft above the thickest part of the atmosphere allowed a smaller rocket to put a payload into orbit, compared to a ground-launched missile. These launches placed satellites in Earth orbit in mid-1958 and, although secret at the time, represented the fourth and fifth successful U.S. satellites, within a year of the Russian-launched Sputnik.

By 1960, Howard Wilcox had moved on to become head of Research and Engineering at General Motors. The space program was just getting under way. In addition to the manned Mercury program, robotic probes were being designed to travel to the Moon. In 1962, JPL issued a Request For Proposal for a high-mobility vehicle intended to operate on the surface of the Moon. Wilcox had recently hired a sharp mechanical engineer, M. G. "Greg" Bekker, who had a flair for issues of vehicle-terrain interaction and mobility design. Wilcox turned to Bekker's team to respond to the announcement.

Bekker was up to the task. With internal company funding, he and his team built a six-wheeled vehicle with balloon tires that at first glance looked like it was all wheels. They made a movie of the vehicle driving through a daunting desert obstacle course with lava flows and boulders. With its melodramatic musical background, future robotics engineers at JPL would refer to the film as *Rover Gladiators on the Moon*. However, the mobility capability of the vehicle was impressive. It looked like it could go over almost anything. General Motors won the contract to build, evaluate, and deliver two prototype Surveyor Lunar Rover Vehicles (SLRVs) to JPL by 1965.

The SLRV prototypes were to be smaller than the original concept vehicle. They needed to be able to fit into the already-designed Surveyor lunar lander. Another requirement imposed on the prototypes was low weight, so that they would not exceed the payload carrying capacity of the Surveyor, and would require less of the scarce onboard battery power to operate. The vehicles were each given a single TV camera to allow a remote operator to see the terrain ahead of the rover and so safely drive it across simulated lunar terrains.

In 1964 there was an open house at the General Motors Santa Barbara facility. Among other displays and demonstrations was one about lunar rovers. The SLRVs had not yet been delivered to JPL, so visitors to the open house, largely family and friends of General Motors employees, were given the opportunity to drive one of the SLRVs on a test track that could be seen only through the rover's own camera. The track was a rugged "moonscape" hidden on the other side of the building from where the open house demonstration was set up.

Brian Wilcox, barely out of elementary school, drove the rover that his father's team had built. The control box let him drive the vehicle forward or backward a fixed distance, stop, and turn left or right. Adjusting the controls, he could put the SLRV into a shallow, medium, or sharp turn angle. Then, by hitting the "forward" button, he would make the rover drive four feet or so through the arc defined by the steering angle. By watching the images from the onboard camera, Wilcox found it easy to keep the SLRV between the small rocks that defined the bounds of the "safe" path.

But there was another mode the engineers had rigged up for driving the rover: They delayed the display of the video images by a few seconds to simulate the time it would take for those images to arrive if they were coming all the way from the Moon, which was where the final version of the SLRV was intended to go. The Moon was 240,000 miles away, so even traveling at the speed of light, the pictures from the rover would take about a second and a quarter to reach the Earth, or two and a half seconds for a round-trip message. Even worse for would-be drivers, the demo designers set the video feed to shut off whenever the rover was in motion. This also was realistic, because a rover driving across the Moon wouldn't be able to keep the antenna of its video transmitter pointed at the Earth except when it came to a stop. If you drove the SLRV in this mode, it was like sitting behind the wheel of a car, looking where you wanted to go, and then closing your eyes while you put your foot on the gas; after you braked, you opened your eyes again to see where you had ended up.

When Wilcox tried to drive the SLRV in its simulated lunar mode, he got into trouble. Without visual feedback from the rover as it was moving, he couldn't tell exactly how the rover got to where it ended up. Since he couldn't see the rover itself, but only the terrain in front of the rover, he had no cues to remind him whether the rover was in the middle of a turn or aligned facing straight ahead. The video image might show a clear path apparently directly in front. But when he commanded the rover to drive its set distance ahead, the next image that appeared would show a completely different view, with none of the rocks he'd last seen anywhere in sight. The rover had actually turned ninety degrees to the right, and was looking off the edge of the safe path into treacherous terrain. But the video information wasn't sufficient to tell him that that was what had actually happened. Unless he could remember all of the commanded motions and turns he had used previously, he wouldn't know whether he had turned left or right. Which choice should he make to correct the problem? He wasn't sure. So he guessed, and often guessed wrong. The attempted corrections could become wilder and wilder, until Brian (and everyone else who tried to drive the video-impaired and time-delayed rover) failed to keep the SLRV within its test track. Of course, if the engineers turned the continuous video feed back on, driving the SLRV suddenly got easy again.

Brian Wilcox learned a lesson that day. The continuous video was giving him huge amounts of useful information he hadn't appreciated before. Without that feedback, you just couldn't drive.

Soon after that open house, the General Motors SLRV was delivered to JPL.

✳

Howard Wilcox was trained as a physicist, and he began teaching his son the science early on. Brian took to the subject. "I was good at it because I was tutored in it. And I found an amazing power that came from being able to analyze things." He majored in physics and mathematics at UC Santa Barbara, attaining dual degrees, and was accepted into the UC Berkeley graduate physics program, the premier such program in the country. But after the first term at Berkeley, he'd had enough. "You can do practical things with undergraduate physics. But graduate level physics . . ." Brian went back to a company where he had worked during college, and eventually started a couple of companies of his own. One of those companies, Polymorphic Systems, produced one of the first microcomputers for the computer hobbyist, the Poly-88. The company did well, selling hundreds of units per month in 1977. Then another startup, named Apple Computer, released its own computer, and Polymorphic Systems was doomed.

It was time to find something else to do. Brian Wilcox had been interested in robotics for some time, and he decided he wanted to be in the robotics business. "My background was physics, my application had been computers, so it seemed natural that computer control of mechanical systems was the place that my interests were best used, in particular microprocessor systems." So he began looking for opportunities in robotics. Helpful information came from a surprising source: his father. Howard

Wilcox had worked on a number of projects since his days at GM. One of the recent projects had been a survey of people and institutions doing research and development in the field of robotics. Howard Wilcox gave his son a copy of the study, and the younger Wilcox sent out résumés. One of those résumés went to Carl Ruoff at the Jet Propulsion Laboratory.

The day Brian Wilcox's résumé reached Ruoff's hands, Ruoff had just been told by his boss to find more "well-educated but entrepreneurial" people to hire. Brian was just what he needed.

✳

Once the GM SLRV was delivered to JPL in the mid-1960s, engineers there drove the rover, evaluating its performance in the Arroyo Seco. But the program to send a robotic rover to the Moon before the Apollo landings fizzled out. The SLRV went into storage. Sometime in 1972, Howard Primus, a technician in the Automation and Control section, took possession of the rover. No one else seemed to think the rover was good for anything, except taking up storage space. Whenever they came across the rover, Primus's managers would tell him to have it hauled away. Instead, Howard would move the SLRV to a different storage place, keeping the rover out of sight as much as possible. Months would go by, then a supervisor would stumble across the SLRV. "Get rid of that thing!" he would say. Primus would just move the rover again. He just couldn't see disposing of such a unique machine; surely, someday it would find a new purpose. "Howard just loved that rover," Carl Ruoff would later remark.

Howard Primus's shell game went on for about ten years. By 1982, Ruoff was supervisor of the Robotics group. One day Ruoff got a call telling him that the storage trailers outside of his research laboratory needed to be removed. The trailers contained overflow materials from the teleoperations lab, so that made them Ruoff's responsibility. They sat in a narrow outdoor parking lot just to the south of Building 198, and the division manager wanted that space for additional parking. So Ruoff got the key to the trailers and went to see just what was stored in them. He opened up one of the trailers, and "By God, there's this white rover. So I went to get Brian [Wilcox] and said, 'Hey, Brian, I think we've got a rover here. Let's take a look at it.' So Brian and I went and opened this trailer [again], and he says, 'You know, I think that's the rover I drove when I was twelve years old, and I drove it into a ditch. We can't really get rid of this rover.'" Ruoff wanted to restore the SLRV to working order and restart robotic vehicle research at JPL. He had Howard Primus bring the vehicle into the basement laboratory area. Ruoff's instructions to Brian Wilcox were simple: Make it work again.

✳

Ruoff needed funding and knew of possible military uses for robotic vehicles. The U.S. Army was getting interested in the idea of a robotic reconnaissance vehicle, one that could go places where soldiers were unlikely to survive. Ruoff went to the JPL program office representative for military-funded programs and asked for money to make the SLRV operational. The manager gave him five thousand dollars.

Ruoff, Wilcox, and Primus went to work. Surprisingly enough, once they had the go-ahead, it took only a few days to get the SLRV rover functioning. The vehicle needed replacement rechargeable batteries and new wheels. Primus had carefully hoarded spare parts for all of those years the SLRV had been otherwise ignored. A spare battery set was among them, good enough until a new set could be identified, ordered, and installed. The wheels were another matter: The wire-mesh, rubber-covered wheels had been an expensive specialized design in the early 1960s, intended to possess many of the characteristics of the final designed-for-flight wheels. (Although the rubber covering would not have survived the temperature variations of the lunar surface.) The wire-mesh wheels had been very lightweight, which was important in enabling the machine to perform in uneven terrains. If the vehicle were too heavy, it just wouldn't be able to move under its own power. There was no way they could afford to duplicate the flightlike wheels.

But then again, they didn't have to. What Ruoff and Wilcox wanted to do with the SLRV was to create a working platform for robotic vehicle research, one that would allow them to experiment with control strategies and maybe onboard robotic "intelligence." If the rover could move reasonably well, that would be good enough. Maybe plain old All-Terrain-Vehicle wheels would do the trick. So Ruoff and Primus drove over to the

Honda motorcycle dealer in Pasadena and took a look at the ATVs there. Some of the wheels looked like just about the right size. But the wheels were steel and seemed to be too heavy for the motors on the SLRV to handle. Next, Primus went to a dune buggy store, and found some wheels that would do the job. Once he had the wheels in hand, Primus machined six adapters to mount a wheel hub to each of the rover's drive motors. To reduce the weight of the wheels further, Ruoff asked Primus to grind the treads off the tires, making the rubber as thin as possible while maintaining its structural integrity. Since the SLRV drove so slowly and weighed so little, the tires didn't really even need to be inflated. At a top speed of half a mile or so per hour, there was little risk of the tires falling off the wheel rims.

＊

When Wilcox had first come to work at JPL, he had been assigned to Ruoff's study team that was jointly funded by the U.S. Army Engineer Topographic Laboratories (ETL) and the Defense Advanced Research Projects Agency (DARPA). The purpose of the study was to identify the research necessary to create robotic reconnaissance vehicles useful to the military. The team focused on how to direct a rover through its immediate surroundings and avoid the hazards in its path. The long-term objective was to produce autonomous vehicles that required almost no human intervention. Give the robot a goal, miles away. It finds its own way there, even avoiding the enemy if necessary. In the short term, coming up with a way to intelligently command a teleoperated vehicle was a more realistic objective.

The challenges of directing a robotic vehicle on the battlefield were not so different from operating a rover exploring another planet. You couldn't afford to communicate with a rover on Mars very often: There was no way to give a rover enough power to send a strong enough signal the millions of miles back to Earth for continuous video. If a battlefield reconnaissance robot sent a strong signal too frequently, the enemy could track it and destroy the robot. So in both cases, whoever was operating the vehicle needed to be able to do it with very limited information.

Now that they had the SLRV rover, Wilcox and JPL were in a better position to market robotic vehicle technologies to the U.S. military. The SLRV was a platform on which to implement and prove those technologies without starting from scratch. And Wilcox had a navigation concept he had first envisioned during the ETL study, one he called "stereo waypoint designation."

If you can't afford to watch video coming back from a camera mounted on the rover, what can you do? You still needed to know what was around the rover, what the terrain looked like. Well, the other extreme from a live TV picture would be just a single frozen picture or one video frame. But that wasn't quite enough. With just one picture, a close-up pebble might take on the appearance of a more distant boulder. Some terrains would appear flat in a photograph, but be obviously undulating to a person standing next to the rover. What you needed was two eyes on the rover: stereo cameras to provide a 3-D view of the terrain. Most predatory creatures with two eyes—humans included—use binocular vision to estimate distance to objects in the environment around them. With two cameras viewing the same object, it was possible to determine the range of the object by triangulation. The line connecting the centers of the two camera lenses is the base of the triangle; the lines between the cameras and the object form the other two sides of the triangle. If the cameras were mounted side-by-side a fixed distance apart, the known geometry would enable range determination. People were also natural experts at doing something that no one yet had a very good way of programming a computer to do: looking into a 3-D display of a rock-strewn terrain and sensing immediately which areas were safe and which were to be avoided.

Wilcox's insight was that a stereo vision display could also be used as a command input device. If he could design a 3-D cursor, operated by a joystick and displayed in the middle of the 3-D terrain, then a person could move the cursor until it appeared to be located at a good safe target location for the rover. If a person designated several of these target points in the display, they would constitute a safe path. If the system was properly calibrated, the target points in the display could automatically be converted to target coordinates, and then into a series of motion commands for the vehicle. The commands could then be transmitted to the rover, which would then execute them, driving a safe path through the intermediate points along the way—waypoints—to the final destination. Once the rover reached its commanded destination, it would take a new pair of stereo images and send them back to the operator. Then the whole process would be repeated until the rover achieved its ultimate destination.

Wilcox proposed this concept as a new vehicle control technology that JPL could provide, one that could be tested out using the restored SLRV. In mid-1984, Wilcox and Joe Hanson, one of the JPL points-of-contact to potential Army customers, made a marketing trip to the U.S. Army Tank Automotive Command (TACOM) to try to sell the idea, with no immediate success. On the plane flight home, Wilcox and Hanson discussed what they would need to do to get TACOM to bite. The problem was that you almost had to have a working version first. Wilcox was convinced that it was doable. Hanson thought it was a bigger job than Wilcox imagined. While Wilcox and Hanson talked, they had a couple of drinks. Before the airplane landed in Los Angeles, Wilcox boasted that he could personally get stereo-image-based target designation up and running on the SLRV, working only a couple of weeks of evenings and weekends. The implementation would be crude, rough, and inaccurate, but at least it would be a concept demonstration.

Hanson didn't believe it could be done. He was almost right.

*

First Wilcox needed to mount a pair of video cameras on the SLRV. For stereo to work, the cameras had to be mounted side by side an appropriate distance apart. The cameras should be aligned so that an object viewed by both cameras would be at the same vertical position in each camera's view. Otherwise, it would be like a person with one eye looking at the ceiling and the other pointed toward the floor. The spare cameras available in the robotics lab had lenses that gave them a thirty-degree field of view. Ideally, each camera would have an optical mount that would allow each type of camera rotation (pan, tilt, roll) to be independently adjusted, enabling the pair of cameras to be carefully aligned with each other. Wilcox didn't have the money or the time to buy or build a complex mount, so he drilled holes in a simple aluminum bar and mounted the cameras about twenty inches apart. He then affixed the bar to the top of the front compartment of the SLRV, so that the cameras faced forward.

Now Wilcox had to get his hands on a stereo vision display. He knew that the teleoperations group in the basement of Building 198 used 3-D displays in some of their work and got permission to borrow one. The system was not as impressive as one might imagine. It was mostly just a steel frame in which two TV monitors had been mounted at right angles to each other, one at eye-level facing the person viewing, the other monitor above and forward of the first, but facing down toward the floor. A half-silvered mirror was mounted at a forty-five-degree angle between where the two monitors faced. In front of each screen was a polarizing filter. The result would be that a person looking into the display would see the images from both monitors superimposed: the first TV could be seen, somewhat dimmed, through the half-silvered mirror; the other screen was visible as a reflection in the mirror. Of course, the mirror-image of the second monitor was just that: backwards. To correct for this, one wire inside the TV had to be resoldered, reversing the scan on the CRT and causing the monitor itself to display the video feed backwards. In this case, two wrongs did make a right, and the reflected backward image was left-side-left and right-side-right.

To achieve the three-dimensional display effect, a person would don a pair of polarized glasses, like those worn by movie audiences for 3-D movies in the 1950s. The polarizing filters in the glasses would allow through only light that had been polarized in the same direction as the filter. The polarization on the left side of the glasses matched that of the "left" monitor in the stereo display, while the right side of the glasses matched the "right" monitor. So each eye would see an image from a different TV monitor. Properly hooked up, this meant that the person looking into the display would see the image from the left SLRV camera with his or her left eye and the image from the right camera with his or her right eye. This created a sense of depth and understanding of the scene ahead of the rover that no single image could have supplied.

A real stereo designation system would have "frame grabbers," electronics that would capture a single video frame from each of the video

cameras on board the rover, so that only two images (one from the left camera and one from the right) would be sent back to the control station at a
time. But Wilcox had no computer or frame grabbers on the SLRV. So he
ran fifty feet of coax video cables back from the SLRV cameras directly into
the stereo display. For purposes of the first demonstration, live video would
substitute for the frozen images that the real system would depend on.

Wilcox now had a 3-D viewer that showed him what the rover saw. He
still needed a tool for telling the rover what to do: a 3-D crosshairs and a
means of moving it around in the display. This would give him a way to
designate points in the stereo image. There was a custom image processor
in the lab that had circuitry that counted the rows and columns of images
as they came out of the camera. Wilcox designed and built a simple modification to the processor to make a stereoscopic crosshairs overlay on the
stereo display. It looked like a big white + sign constructed of horizontal
and vertical bars drawn into the two camera images. The crosshairs had
the flexibility to be shifted not only left-right and up-down in the display,
but also in-out, representing the distance out in front of the rover. To do
that required separate left-eye and right-eye crosshairs: While the horizontal bars of the two crosshairs would always be superimposed on each
other, the vertical bars would be offset from one another. This offset
would create a perception of depth to the human viewer. Adjusting the
offset would make the crosshairs move directly away from or toward the
viewer. Wilcox bought a joystick from the local electronics surplus house.
After modifying and interfacing the joystick to the image processing system, he could control all three directions of crosshairs motion. Appropriate software could convert the 3-D crosshairs position on the screen into a
target location to which the rover could be directed to drive. With more
software, the computer could generate output signals to turn on any or all
of the eight motors (six drive motors and two steering motors) on the
SLRV. Wilcox ran more wires to carry the output signals from the computer to newly installed analog amplifiers on the SLRV. Now for the first
time he had an off-board computer capable of driving the rover.

Wilcox went to the JPL materials supply in search of paint. Only a few
colors were available. He called one of the JPL in-house video producers,

and asked her which choice would look best to the camera. The answer came back: blue. So Wilcox selected the only blue paint from among his options and gave the SLRV its first new coat of paint in twenty years.

Wilcox was running out of time. As he worked to develop the first version of stereo target designation, he found himself devoting more of his "evenings and weekends" than he had originally imagined when he had made the bet with Hanson.

Wilcox's first implementation of a rover driving algorithm was very crude. There were no sensors on the motors to measure how far the rover had actually driven or turned. The computer program that operated the rover's motors just assumed that the rover's speed was one foot per second. So if the target location were two feet ahead, the software would run the motors for two seconds. If the destination were twenty feet away, it would run the motors for about twenty seconds, and so on. But the real rover was more variable: It would slow down when going uphill, speed up on the downhill; and the wheels could slip on sandy surfaces, so that the vehicle covered less distance than it would ideally. The steering motors on Wilcox's rover were run by timing just like the wheels. To get a maximum turn, you just ran the motors long enough to be sure they had hit their stops. Going straight was harder, since it involved running each motor just the right amount of time away from the turn limit. The time it took to turn was different for front and back, and for left and right. That was one major complication that Wilcox's "quick and dirty" software was not intended to handle. Just as the drive motors were assumed to roll the vehicle forward at a certain distance per second, so the steering motors were expected to turn the front and rear vehicle compartments through a certain angle per unit time. When the rover completed its commanded turn, it would usually be pointed only in the general direction of its target, so that as it drove forward it drifted significantly to one side of its intended destination.

But, after a fashion, it worked! Wilcox could look into the stereo display, adjust the 3-D crosshairs using the joystick, and hit the button to choose a target. If he wanted to, he could select several targets, creating a multi-segment path, but he would then have to remember where the earlier targets were, since he didn't have any way to display them on the screen. That would come later. He'd done what he'd promised on the airplane.

When Joe Hanson saw the system, he was floored.

✳

Wilcox's first effort did not work well enough to ever give a live demonstration, but the basic functionality was there. Enough that the rover and the control station could be used to explain the basic elements of the system. Wilcox made a homegrown video that went through each step of the process: The rover transmits stereo pictures back to the control station; the operator designates a target position in those images; commands are sent to the vehicle; and the rover goes there. On the strength of that tape and further discussions, TACOM funded a new JPL task starting in late 1984 to develop and demonstrate a stereo-waypoint designation system.

The new system filled a niche between a remotely driven or teleoperated vehicle and the holy grail of the fully autonomous robotic vehicle that needed no guidance at all from human operators in order to complete its navigation task. A teleoperated rover was just being remotely driven, with the onboard communications system allowing the controls and human driver to be located somewhere else. The new system was a step up from this. Charley Beaudette, the TACOM technical director for the task, coined a new term: Computer-Aided Remote Driving, or CARD. More than twelve years later, when Sojourner rolled across the dust of Ares Vallis toward rocks named after cartoon characters, it would be a variation of CARD that would guide her.

But there was yet a long path to follow to prove CARD a practical technology.

FOUR

THE RIGHT PLACE AT
THE RIGHT TIME

The two projects Ed Kan and Neville Marzwell had hired me to do were coming to an end.

I was successfully completing my JPL assignments, yet a nagging question remained: What was I going to work on next? I started to worry. I was learning that projects and tasks at JPL had beginnings, middles, and ends. Those tasks provided thirteen-digit charge numbers. And if you didn't have an authorized charge number to put on your time card, you could be out of a job. When Ed Kan first offered me the job at JPL, he had been noncommittal about future employment, only indicating that it was likely that other things would come up, if I did well.

It turned out that JPL had a special charge number for employees who had no real tasks to charge to. It was referred to as "090" for the number of the artificial organization within JPL that administered it. If someone was out of work at JPL, they were "on oh-nine-oh" until they either found work or were laid off. During the time you were charging to 090, your job was to find a job. JPL was generous: Each employee was entitled to one month plus one week of 090 charging for every year he or she had worked at JPL.

No one wanted to get on oh-nine-oh.

I was still a JPL rookie, and I wondered where my next job was coming from. Yet opportunities did present themselves. One of the engineers in the building next door was the task manager for a Space Shuttle flight experiment. The intent of the experiment was to build and test a force-torque sensor to be installed on the shuttle's manipulator arm. I was astonished to learn that the huge Space Shuttle manipulator, regularly used to lift large satellites out of the cargo bay and deploy them into orbit, had no sensors on it to tell when it made contact with an object. For the astronauts controlling the Remote Manipulator System, it was like trying to pick up delicate pieces of machinery with an arm shot full of Novocain. The only protection against knocking the mechanical arm against the side of the shuttle was the vigilance of the astronauts as they watched through windows or TV cameras looking into the cargo bay. The force-torque sensor was to be mounted at the "wrist" of the manipulator. It would communicate back to the astronaut operating the arm the direction and magnitude of the forces the arm was generating as it encountered objects in the cargo bay. With that kind of feedback, things were less likely to get broken and much more delicate work would be possible. The task manager asked me to be the system engineer for the project. Well, that was the first new job I was offered. It seemed like a good one: a chance to work on something that would actually fly in space.

And then one day Brian Wilcox and Carl Ruoff asked to meet with me. I didn't know either of them well. I knew that Ruoff had been the supervisor for the Robotics group, and that Wilcox had taken over the position when Ruoff went on to other things. Ruoff would pass through the building once in a while, ask questions about what we were working on, and start conversations about engineering philosophy.

The Robotics Laboratory was in Building 107, and most of the engineers in Wilcox's group had their offices there. I sometimes found myself in lunchtime discussions with Wilcox. He had a sarcastic wit that was always active but never vindictive. I learned that I couldn't tell when he was putting me on, so it was a safe bet that if it occurred to me that he might be joking, then he was.

I sat in the Building 107 conference room facing Wilcox and Ruoff. One of them said, "We notice that when you work on a job, things happen. Things get done." They had a task that was languishing. It needed someone to prod it into producing results. They described to me the "Blue Rover DDF."

DDF stood for Director's Discretionary Fund. Compared to most large companies, JPL had virtually no moneys devoted to self-directed research and development. The DDF was what little JPL had of this type of funding, which the director was free to allocate as he or she saw fit. In practice, a committee within JPL reviewed proposals submitted by employees, awarding amounts of $10,000 or $20,000 for the most promising ideas. A couple of years before my meeting with Wilcox and Ruoff, Carl had succeeded in selling the first big DDF: a several-hundred-thousand-dollar, multi-year effort to develop technologies for planetary rovers. The principal selling point for the research was that these technologies were crucial to the future of the Laboratory: Most of the planets of the solar system had already been explored by JPL spacecraft, or would soon be visited by the Voyager mission currently in progress. Once the current phase of deep space exploration was complete, the next step would be to explore the surfaces of the planets and moons in the solar system. Unless JPL was ready to take up the challenge when the time came, the Lab's future funding, and indeed its reason to exist, could be in doubt.

As originally funded, the DDF was far reaching in terms of the research questions to be addressed. How would a robotic arm mounted to a mobile platform pick up samples? How would the rover determine its location in the terrain using its stereo cameras? How would the rover recognize hazards? And how would it automatically plan a path around those obstacles to the goal? The time was approaching when the DDF would need to demonstrate tangible answers to some of these questions.

There were several research engineers participating on the DDF, mostly in our section, but also a few in Section 366, where AI-type software was supposed to be done at JPL. It was important to spread the work around some within the organization. This was the biggest DDF ever up to this point, and it gobbled up a lot of the funds that could have gone to other sections.

Ruoff and Wilcox were asking me to lead a team (albeit a small, poorly funded, part-time one) that was signed up to make the Blue Rover smart enough to drive itself to destinations we specified, while avoiding hazards along the way instead of driving into them. Once this became clear, I realized that I couldn't imagine a better job. The whole task was just infused with a sense of turning science fiction into reality. And the tools to do it had already been mostly assembled. All that had to be done was to bring the pieces together.

✳

The crux of the Blue Rover DDF was "Semi-Autonomous Navigation." The name was dull and yet—to my engineering soul—magical: Hidden inside was a combination of technologies that would enable a robotic vehicle to drive miles on its own across an alien landscape without help from its human masters.

SAN depended on a repeated sequence of activities: sense-perceive-plan-act. Many of the functions in CARD that were done manually would instead be performed by sensors and software. The rover would now be required to sense its surroundings, transform the raw sensor data into an understanding of the safe areas and hazards in the vicinity, plan its path through the terrain, and finally execute the plan. The rover would reach nearly to the limits of what it had been able to sense, then stop. For SAN to be useful, the vehicle would have to repeat the entire process over and over, effectively putting one foot in front of the other, building up the yards of traverse. If the rover could move only a few yards per minute on its own, it could cover many times the distance allowed by the simpler CARD approach in the face of a forty-minute round-trip communications time delay between Earth and Mars.

Many of the mission concepts circulating at the time proposed a Mars orbiter that would take high-resolution pictures of the surface, producing maps of large areas distinguishing terrain variations of several feet. Since the orbiter would be there anyway, rover operators on Earth could take advantage of these maps to plan a "global" route for the vehicle, one that extended far beyond what could be seen from the rover's onboard cameras. If the orbiter maps were extensive enough, perhaps rover routes several miles long could be planned.

The map for the terrain surrounding the rover, together with the plan for the route through that terrain, would be uplinked to the vehicle. At this point, the rover would be on its own. Just like a CARD rover, it would capture a set of images with its stereo cameras. But instead of sending these pictures back to a human operator, onboard software would process the images to generate a "local" map of the nearby terrain. "Machine vision" software would compare points in the left and right image views and determine the distances to features in the scene. These distances would then be transformed into an overhead map. Due to the vantage point of the original images, there would be much detail in the portion of the map close to the rover, and less farther away. If nearby rocks blocked the rover's view, there would be "shadows" in the map where no data was available at all.

The local map would now be matched against the global map sent from Earth. "Terrain matching" the two maps was like placing a tracing-paper drawing of one map down on the other, and then sliding the maps back and forth until they were aligned. Once you got them lined up, you could merge the two of them together into a single map that was better than either of them alone: The local map gave you much higher resolution in the immediate vicinity, and the global map gave you coverage beyond the rover's sight. And once the maps were merged, the rover would know exactly where it was positioned relative to the global route plan it was trying to follow.

The merged map would be analyzed for traversability. If the map elevation changed too much from one map point to the next, this was a good sign of either a steep slope or a big rock. Without really understanding what a rock was, except as a collection of higher-elevation points surrounded by lower-elevation ground, the software could divide up the map into hazardous and safe regions.

Here the automatic path planner would kick in. The planner looked for paths through the traversable parts of the map to waypoints along the global route. Once it found one, it would pass it along to the rover motion-control software, which, just as in the CARD system, would drive the rover to the waypoint.

If you could just do this over and over, without error, you'd be home free.

✳

Navigating a physical vehicle in the real world forced the team to deal with error. This was not really error in the sense of a mistake, but error in the form of measurement uncertainty.

Ideally, you could tell a rover exactly what path to take to where you wanted it to go, and it would perfectly execute that path, arriving at the destination, however distant, like a car pulling into a parking space. But even well-designed sensors have limitations in the precision with which they can take measurements. If the magnetic compass that told the Blue Rover the direction it faced was off by just one degree, the rover could have drifted two feet off its course by the time it had traveled one hundred feet. The counters on the six rover wheels could measure wheel rotation to within two inches, but the wheels themselves could slip in sand or on rocks. All wheels didn't cover the same distance: If one wheel went over a rock, and another rolled over flat ground, the one negotiating the rock had a longer distance to cross to keep up with the other. If the terrain was rough, the rover might cover more than a hundred feet to get to a spot ninety feet distant. So odometry measurements were inherently uncertain. The result was that rovers never went exactly where they were supposed to go.

The deviation between a vehicle's planned path and the path it actually follows is called "dead reckoning error."

Why don't the same errors apply when you drive a car, or even just walk down the street? If you closed your eyes after picking your destination, and tried to get there blind, the same problems would show up. But a human driver is constantly correcting for the difference between where the car is going and where the driver intends for it to go.

Wilcox knew dead reckoning error would be a problem, and had planned to implement "visual tracking" software to compensate for it. For the Blue Rover, all paths were defined as a combination of turn angles and straight-line path segments the vehicle would execute. The visual tracker would use images from the rover's stereo cameras to improve the rover's execution of both turns and straight-line traverses. The tracker compared successive images. During turns, objects seen by the cameras would move either from left to right, or right to left through the images, depending on the direction of the turn. How far an object moved across the images was a good measure of how far the rover had turned.

For straight-line forward motion, the tracker would compare "windows" within successive camera images. A single full TV image for the cameras on the Blue Rover was composed of dots, or "pixels," arranged in an array 320 pixels across by 240 pixels up and down, like a pointillist painting. The tracker focused only on a small window, 30 pixels by 20, centered on the expected end point of the straight traverse. When the next image came in, the original window was checked against a window in the corresponding region of the new image, and against new windows shifted just slightly left, right, up, and down. Whichever window matched best with the original window identified the direction in which the rover had drifted off course, and the rover could now improve its dead reckoning. The best-matched window in the new image became the "original" window for the next round of visual tracking.

Wilcox needed to upgrade the stereo camera mount he had originally constructed so hastily. The cameras needed to be isolated from the vibration induced when the Blue Rover rolled over rocks. Otherwise, if the cameras jerked around too much, the visual tracker would lose lock and become useless for correcting the rover's dead reckoning error. Wilcox designed a new camera mount that allowed the cameras to pivot in any direction, and returned them to upright orientation using a set of springs. But he still had to smooth out the camera motion enough to stay within the inherent limitations of the visual tracker. He came up with a scheme that depended on wide paddles moving through a thick liquid, like molasses. The paddles would be fixed to the camera mount, and jars of the molasses-like stuff would be mounted to the rover body.

Wilcox stopped by the drugstore and bought several infant-sized baby bottles and a can of STP oil treatment, which was a black, highly viscous liquid. The baby bottles would be perfect! He could fill them with STP and use the nipples to seal the oil in so it didn't get all over everything. "I got some really odd looks at the checkout counter." Apparently the checker wondered just what Wilcox was going to be feeding his baby . . . "The cashier said, 'That should fix the little sucker.' Really." The other engineers working on the Blue Rover would also tease Wilcox over his unorthodox choice of off-the-shelf components. But the baby bottles did the job. "The camera mount stabilized the images just like a good cameraman as compared to the jerky motion of a novice with a new camcorder. And the visual feature tracker didn't have to search more than one pixel per frame for the motion."

The rest of the Blue Rover upgrades were falling into place. A microcomputer had been mounted in the rover's middle compartment to control all of the onboard motors. There were now image buffers installed to freeze images from the rover cameras, as a true CARD system required. We still could not afford to install the buffers on the vehicle itself, so the live video from the cameras had to be transferred over a long cable back to the control station.

To test the new driving algorithms in the lab, the rover was suspended from the ceiling by braided steel cables. S-hooks on the ends of the cables slipped through rings bolted to either side of each compartment, six in all. The rover would hang in the air, wheels turning, driving but not getting anywhere. When it attempted a turn, the vehicle would twist in its restraints, drive its forward and rear wheels, then twist back. The suspended rover was vaguely reminiscent of a cross between a puppet on a string and some kind of medieval torture device.

***

"Hard right. Hard right . . . No! Hard left!" Wilcox's voice came over the walkie-talkie. Brian Cooper rolled his eyes. He was the engineer upgrading Wilcox's jerry-rigged control station—but right now he was driving the Blue Rover in response to Wilcox's commands radioed from the field. Cooper swung the joystick over, and waited to hear the next instructions over the walkie-talkie. After a few seconds, the images on the monitor in front of him panned wildly. "Stop! Stop!" Cooper complied, hitting the red emergency button. He shook his head. Brian Wilcox was a brilliant engineer and a good supervisor, but he was lousy at directing the movements of the Blue Rover to a remote operator.

They were trying to move the rover into its starting position for CARD testing in the Arroyo Seco. Getting the rover there was a production in itself. The Blue Rover would start out suspended in the air from six steel wires for in-laboratory testing. First a couple of engineers would drop the rover, one wheel at a time, from the wires onto a wheeled wooden pallet. Then they'd push the pallet over to the big garage door on the south face of Building 107, hand operate the roll-up chain until the door had lifted high enough for the rover to pass, and push the pallet outside into the alley beyond. The rover was then carted down the alley and onto Surveyor Road, stopping traffic, rolled past the guard station, through the JPL East Gate, and lifted up onto the sidewalk on the west side of the road-bridge that crossed the Arroyo Seco into the east parking lot. The rover didn't cross the bridge. Instead, once it was safely out of the roadway, the engineers shepherding the rover plugged in the fifteen-hundred-foot umbilical cable that would allow commands and video to pass between the Blue Rover and the rover control station in Building 107. From here, Brian Cooper could now teleoperate the rover at its stately pace down the grassy slope into the arroyo, taking direction from the engineers in the field to ensure that he wasn't putting the vehicle at risk.

Brian Wilcox had first driven the Blue Rover decades before, and understood intimately the issues of time delay that arose when a human operator tried to control a machine thousands or millions of miles distant. But even when the rover was in the arroyo and the driver was only a few hundred yards away in the laboratory, the slow steering of the rover and the lags in verbally relayed instructions made for vehicle response delays as if the Blue Rover were on the Moon. But since Wilcox was out in the field, standing a few feet from the rover, the reality of those delays just wasn't registering. He was just seeing that the rover wasn't going where he was expecting it to, so he had to continuously give new instructions to Brian Cooper. Cooper had learned by now that if he blindly followed Wilcox's directions, the rover's motions would get more and more erratic—just as Wilcox himself had observed long ago during his first experience driving the SLRV. The view from the Blue Rover's camera was limited to a thirty-degree swath, not enough to provide much context for Cooper to teleoperate the vehicle. He could always steer the rover left and right to build up a ninety-degree field of view, but that would take too long, so it *could* be really helpful to get instructions from someone out in the field. All they wanted to do was get the vehicle into position so the real testing could begin. Wilcox was about to direct Cooper to joystick the rover closer to the starting line for the CARD test. Over the walkie-talkie, Cooper spoke first. "Can you put Steve on? I need to ask him something." Steven Katzmann was one of the other engineers participating in the field test. Once Steve got on the line, Cooper was conspiratorial. "Whatever you do, don't give the walkie-talkie back to Wilcox. Now tell me which way to drive."

✳

By the summer of 1986, the Blue Rover team had demonstrated a real CARD system. Brian Cooper could designate a target forty yards in front of the vehicle, and the rover could hit that target within about two yards. That wasn't too bad, and their analysis showed that with higher-resolution cameras they could do better or designate farther. The visual tracking system, even with the aid of STP-filled baby bottles, did not perform so well. It ran too slowly on the computer then available, and was abandoned.

Proving the SAN concept took longer. The thirty-degree field of view of the rover cameras was just too narrow. There wasn't enough information in the output of the stereo machine vision software for the automatic terrain matcher to get a match. So Wilcox took the baby-bottle camera mount off the rover, installed a pan-tilt head, and reattached the cameras onto the head. Now the rover control computer could aim the cameras. By turning the pan-tilt head left, right, and straight ahead, and capturing three sets of stereo pairs, the rover could collect a panorama with a nearly ninety-degree view.

Just as we were getting ready for field tests, our research partner from Division 36 reported that his automatic path planner didn't work. He had developed a clever approach to storing the terrain map at multiple resolutions. But the path planning algorithm he had designed to run on this data had turned out to be a dead end.

Don Gennery, a Ph.D. from Stanford and the engineer in Wilcox's group behind both the stereo vision software and the terrain matcher, had seen journal articles about various planners, and didn't think that path planning should be so difficult. He certainly didn't want to see the failure of what he viewed as a minor element of SAN stand in the way of proving the veracity of his own algorithms and software. So in two weeks he wrote the software code for a working, though computationally intensive planner based on an algorithm he had read about. Like all of Gennery's code, the planner was slow to run. But it worked.

The SAN test runs were excruciatingly slow. Gennery believed in mathematical rigor. He was the group's algorithm master. But he had no interest in optimizing his software for speed. From Gennery's point of view, proving the concept was all-important; everything else was just implementation details. So the machine vision and terrain matching processing took hours. If we were lucky, we could get maybe three SAN cycles in a full day out in the hot dusty arroyo. Members of the team spelled each other in standing guard over the Blue Rover.

In the end, after a lot of literal sweat, the Blue Rover proved the SAN concept, performing a number of sense-perceive-plan-act cycles, and crossing about fifty feet of ground. And a JPL video crew got it on videotape. We had made the Blue Rover navigate in the Arroyo Seco.

*

Rovers were still small change. Every few months, my supervisor would ask if I was ready to work on one of the "big" robotics projects. But it was too late. I had already caught the rover bug. So each time the opportunity came up to work on a robot that rolled rather than reached, or drove to new places instead of being bolted to the floor, I took it.

More money arrived from TACOM, and a new "rover" came with it. The vehicle was a military Humvee, the replacement for the venerable jeep.

Out of the success of the CARD demonstration, TACOM had funded a newer, larger task at JPL, the Robotic Technology Test Vehicle program. The new RTTV team, comprised mostly of engineers who had worked on the Blue Rover, was to reimplement CARD on the Humvee, simultaneously creating a system that could evolve to greater and greater capability.

TACOM also wanted the new system to be demonstrated around the country at Army mobility test courses. So the new control station went into a semitrailer. The trailer was carpeted and air-conditioned (primarily to protect the computer workstations that were located inside). It became the comfortable yet windowless de facto offices for Brian Cooper and Steven Katzmann. When needed, the trailer could be hitched to a tractor-truck and hauled anywhere.

Switching from the Blue Rover to the Humvee created a few new challenges. The Blue Rover drove very slowly. Its paths consisted of a number of straight lines strung together. Whenever the rover changed course, it would come to a stop, steer its three-compartment body into a turn, drive through the turn to its new heading, stop, straighten out, and continue on along a new straight line. The Humvee needed to move faster. No vehicle in motion could turn instantly, so the CARD paths would now have to be curved, and no more sharply than the Humvee could negotiate at speed.

The CARD control station went through a transformation. Commercial technology was now catching up with our needs. We junked the dim 3-D display that relied on polarizing filters, carefully mounted and modified television monitors, half-silvered mirrors, and polarized sunglasses. Silicon Graphics Inc. now manufactured a color graphics computer workstation that could become a stereo vision display. The computer screen could alternate back and forth between displays of two different images, and special battery-powered goggles with liquid-crystal shutters would allow only one eye to see the screen at a time. The goggles were electronically synchronized to the display, shuttering the right eye at the exact instant that the image meant for the left eye was being shown, and vice versa. Even though a person wearing the goggles never saw both images at exactly the same time, the images were updated or "refreshed" so fast—twice as fast as a normal TV picture—that the display was smooth, comfortable to look at, and clearly three-dimensional. The operators didn't get headaches from watching the screen, which had often been a problem with the old system. Alignment of the left-eye and right-eye images was no longer an issue, because both were being shown on a single monitor. There was no way for them to get out of alignment with each other. The total effect was the best 3-D display any of us had ever seen.

Brian Cooper's job was to build the new CARD control station on the new Silicon Graphics workstation. The SGI machine was designed for graphics, so instead of crude crosshairs to designate waypoints, this time Cooper would create a 3-D model of the Humvee. He would be able to turn the graphical Humvee in any direction, and place it anywhere on the 3-D terrain display he wanted to. As he pushed on the joystick to move the model ever farther out into the terrain, the computer-generated Humvee would get smaller and smaller, just like the real thing. You would actually be able to use the model to tell whether the real vehicle would fit between two obstacles, or would have to go around them. And instead of merely marking the positions of waypoints in the images, the control station would display the entire path as a winding road (with yellow bricks!) shrinking into the distance.

Cooper had been an officer in the U.S. Air Force. The Air Force had paid for his college degree and promised to put his engineering skills to good use. He had gotten out as soon as he could once he discovered that he was stuck managing tasks—for which he had no interest—instead of doing real engineering work. When he came to JPL, he soon found himself upgrading the Blue Rover control station. Now he had the opportunity to become an expert at the software tools available for the brand-new SGI machine. It was as if he was getting to design a new 3-D computer game, only this one had the ultimate kid's toy at the other end, a camouflage-painted Humvee. What more could he ask for?

✳

Every year, the Electronics and Control division would conduct a daylong briefing for all of the employees new to the division. The idea was to give the new hires some background on the organization and goings-on of the division. Among the several presentations was a talk, usually given by Brian Wilcox, on "Autonomous Vehicle Research at JPL," which detailed the CARD and SAN technologies we were developing. One year Wilcox told me, "Why don't you do it this time?"

The conference room was filled with fifty or sixty people. As I waited my turn to speak, I watched the presentation before mine. The speaker was referring to a chart that consisted of columns of numbers. The numbers were important—they indicated how much money was coming into JPL for various types of activities—but I knew I wouldn't be able to remember anything from the chart once it disappeared.

I didn't have any charts with numbers.

My turn came. I placed my first viewgraph on the overhead projector. It was a color picture of the surface of Mars, taken by the Viking 1 lander. Reddish-hued rocks were strewn over the landscape, stretching from the foreground into the far distance. "We do research on autonomous vehicles at JPL because we want to go here. And we want to find out what's over that horizon . . ."

FIVE

THE BIG ROVER
THAT NEVER WOULD

I n the last few years of the 1980s, rover research began to heat up. JPL was funded to do a new study for what might be its next big mission: Mars Rover Sample Return, or MRSR. At about the same time, NASA also started up Pathfinder Planetary Rover, a new rover research program designed to develop the technologies that MRSR just might need to succeed.

The centerpiece of the JPL research program was to be the Pathfinder Planetary Rover Navigation Testbed vehicle. As new technologies matured, they would be migrated off of the bench top and onto the testbed for tryouts. The new rover was going to be big, much bigger than the Blue Rover. The increased dimensions of the new vehicle were not driven by any mere showmanship desire for impressive physical size, but by a piece of equipment called the nineteen-inch rack.

The purpose of the rover was to be a testbed for improving and evaluating navigation approaches like Semi-Autonomous Navigation, CARD, and whatever else came next. What the Blue Rover had lacked as a testbed was onboard processing capability: The rover's "smarts" were at the other end of a long cable leading back to Building 107. Future rovers operating on Mars would have to carry their own computing. Proving that all of the computer power necessary to operate a rover could be carried by that rover would be a key feature of the research program. By the same logic, the vehicle would also carry its own power source onboard, capable of powering both the computer and the motors.

The testbed chassis would be loosely based on the Blue Rover design. Like the earlier vehicle, it would have six wheels and three compartments. But one problem with the Blue Rover stemmed from the flexible spring-steel that connected the three bodies. Keeping the rover safe in rough terrain might require knowing the relative orientations of the compartments, so the vehicle could stop before one of the compartments could tip over. There was no easy way to mount a sensor on the spring-steel member, so the new rover would replace it with rotational bearings outfitted with "encoders" that could precisely measure the tilts of the compartments.

What would be the most cost-effective design for a navigation testbed? As Wilcox reasoned, the vehicle was primarily a platform for moving the navigation sensors and computing subsystem around. The rover would be capable of driving into somewhat rough terrain, but only to the extent necessary to fully exercise the navigation software. Since computers and related electronics were improving year by year, it seemed likely that the onboard computer Central Processing Unit would be replaced, perhaps several times, before the mechanical chassis was considered obsolete. The testbed had to allow for easy change-out of image processing boards and new CPUs, as well as other electronic equipment that might eventually be selected. The most readily available standard for electronic equipment mounting was the nineteen-inch rack. One could easily find card cages and power supplies with holes pre-drilled into their faceplates for immediate mounting into these standard-width racks. Two racks would leave plenty of room for expansion.

Once the decision to place two nineteen-inch racks on the rover had been made, the rest of the vehicle began to fall together. The electronic equipment would certainly overheat during outdoor testing in the 90°F+ summer temperatures of Southern California; an air-conditioning unit would be mounted on the back of the racks. The only low-cost portable source of power that could provide the kilowatts needed for the electronics and air-conditioning was a gasoline motor generator, which could be mounted elsewhere on the vehicle. Like the Blue Rover, the testbed would also require a set of stereo cameras to sense the nearby terrain, preferably with an unimpeded view of the surroundings. The electronics rack was already the tallest point on the vehicle, so a pan-tilt head and camera bar would go on top of the rack. And because the research task would eventually want to do something useful when the vehicle successfully navigated to a site, such as pick up a rock sample, an available robotic arm would be affixed to the front. You didn't really want an expensive robot arm to act as the rover's bumper, sticking out in front as the first point of contact between the testbed and a big rock. So the mechanical arm was mounted on a "dunking bird" assembly that would rotate the arm forward when it was needed for sampling, and rotate it back and out of the way during traverses.

The testbed design matured. When it was built, the testbed would be over twelve feet from end to end. We would need six recreation vehicle tires thirty-five inches in diameter to support the vehicle. We went to work requisitioning components, having the chassis elements machined, assembling the pieces. We had a contest to name the testbed, with a bottle of champagne for the winner. Despite the plethora of names submitted, we ended up with "Robby," seemingly the moniker for every robot since the movie *Forbidden Planet*.

❋

Although Robby's size had been based on practical requirements, many people who saw the rover seemed to miss the distinction between a test vehicle and one that would actually be sent to another planet. The most common questions were "Will those rubber tires work on Mars?" and "Do you really expect to send something that big?" We even contemplated spray-painting the tires silver to circumvent some of the inquiries.

The reactions to Robby caught Wilcox off guard. "Many people who saw the Blue Rover said 'Oh, it's so small. I envisioned something bigger.' So I was very surprised at people's reactions to Robby, which was almost exactly twice as big. Their reaction was always 'It's so big.' There must be a narrow band in between which would have generated no reaction at all."

❋

All of the work on Robby was geared to the "One-Hundred-Meter Milestone." JPL had signed up to prove that Robby could autonomously traverse the length of a football field through the natural terrain of the Arroyo Seco.

Once again, the Robotic Vehicles group partnered with the AI-focused Robotic Intelligence group to create the software that would make Robby go. Wilcox's team would develop the sensing and perception software that would build terrain maps from the raw stereo images that would come out of Robby's cameras. They would also do the terrain matching and the actual control of Robby's motors. The Robotic Intelligence group, led by David P. Miller, would write the path planner.

***

Donna Shirley led the MRSR study team. She was one of the few women at JPL to have ascended through the male-dominated ranks of the engineers to a position of high visibility. She had been at the Laboratory since the sixties, working on several flight projects, from Mariner to Cassini. Now she was running the effort that was the first nascent step toward the start of a new mission.

The study team included engineers from throughout the JPL organization: mechanical, power, thermal, telecommunications, electronics, and computing engineers; interplanetary trajectory designers; mission operations people—and rover navigation and control engineers. The study team met weekly. At first Brian Wilcox attended, but he was a group supervisor with too many other responsibilities, so the assignment of representing vehicle navigation fell to me. Don Bickler was my counterpart for rover mobility design.

Shirley clearly enjoyed being in charge, the only woman in a room full of male engineers. Much of the team was young, and Shirley sometimes treated them as a group of children in her charge, requiring her guidance; they were intelligent but naive, creative but needing seasoning.

The objective of the Mars Rover Sample Return mission was to return ten pounds of Martian rock and soil to the Earth. The rover's job in this would be to collect those samples from geologically diverse sites on Mars, then bring them to a lander where a rocket waited that would launch the treasure into Mars orbit and then onward to Earth.

Shirley started the MRSR team out by forcing the engineers into doing "trade studies." She was trying to hold the team back from rushing to
a single design too soon. Often a creative engineer would hit on a clever
idea, then converge on a point design that represented an elegant solution
to the problem he or she had been presented with. Unfortunately, the best
design for an isolated widget might no longer be best in the context of the
entire system.

Shirley made the team consider a range of options for each of the key
technologies. The MRSR rover might use "structured light," laser range
finders, sonar, or stereo vision to see what hazards confronted it. How
many wheels should the rover have? Six seemed like a good number, based
on the Blue Rover and Bickler's work. But why not two, four, or eight?
Maybe the best rover would have no wheels. After all, humans had a lot
more personal experience walking on legs than rolling on wheels. And
much of Earth's terrain was off-limits to wheeled vehicles, reachable only
on foot. The study team members representing each rover subsystem explored the range of design options available in their areas of specialty.

The plan for MRSR was ambitious, and became more so as the study
progressed: The rover was to be a rolling geologist's laboratory. It would
drive for hundreds of miles across the surface of Mars, then be directed by
scientists on Earth to pick up rocks, drill cores, slice, dice, and package
samples for return to Earth. A mapping orbiter flying overhead would create a terrain map for the rover at three-foot resolution; using that map, the
rover would always know exactly where it was on Mars. It would be so
smart that it could drive for miles without guidance from human operators, and might even choose to call home only if it figured out that something it saw was scientifically interesting. MRSR would push the limits of
artificial intelligence, computing power, and robotic navigation. And it
would be heavy, hundreds of pounds. Once it had collected all that it
could hold, the rover would select which samples to keep, and which were
no longer exciting enough to hold on to. The rover would find its way to
the sample-return lander, and transfer over a sample canister with several
pounds of rock and soil. Finally, the return vehicle would launch itself
back to Earth with its processed alien cargo.

❋

Robby was not to be the only focus of the Pathfinder Planetary Rover program.

Don Bickler finally got the money to build mobility models. He and a few of his team of mechanical engineers wanted to prove that the mobility performance of the small models would be "scale-invariant." If the rock-climbing and stability characteristics of tabletop rovers could be directly extrapolated to their full-sized versions, then the mechanical engineers would have a relatively cheap and easy approach to trying out new mobility concepts. It seemed plausible, but before they accepted the notion, they wanted to prove it by example. So in the summer of 1989 Bickler's guys built Robby Jr., a one-seventh scale version of Robby. They would come into work in the morning, then spend the afternoons in Bickler's garage machining the rover components. Among them was Howard Eisen, a co-op student from M.I.T.

Eisen made the study of the two Robbys the subject of his Master's thesis. He formulated mobility tests, then conducted them easily on Robby Jr., then with more difficulty on Robby. Eisen had huge aluminum rails delivered to 107. He set them up at several different angles, creating slopes of various tilts up which Robby would drive. The work proved scale-invariance, and incidentally showed that Robby's existing wheel drives were too weak for all but modest traverses in outdoor terrain. Upgrading the torque capacity of the wheel drives to extend the testbed's operating environment became a new milestone. Eisen went back to M.I.T. to complete his thesis.

Since building the pantograph rover model, Bickler had continued to toy with vehicle concepts. He wasn't satisfied with the trouble the pantograph—or the Blue Rover—could get into when attempting to drive over a "bump" rather than a "step." Bickler wasn't crazy about computers or software, and liked to rile up Wilcox and his group by saying that a good mobility system wouldn't need any of "that onboard intelligence stuff." But sometimes computers did have their uses. Bickler found himself tinkering with designs by computer analysis, rather than building every

minute variation he came up with. One of the six-wheeled designs he
tried was much simpler than the pantograph, with a smaller bogie at one
end of a larger master bogie or "rocker." He played with the proportions
of the bogies and the sizes of the wheels. The vehicle's hazard-crossing capability seemed highly dependent on these proportions. Eventually Bickler hit upon a version that the computer analysis said would outperform
the pantograph.

In November 1989 Bickler's garage produced another rover model,
with Bickler's crew of young engineers providing much of the labor.
Imagining Robby to be a full-sized Mars rover, they sized the "rocker-
bogie" to be a one-eighth scale model. The rocker-bogie rover worked as
the computer analysis had predicted, climbing as well as the pantograph
over steps, and surpassing the pantograph when driving over bumps. Simpler and better! They really had something now!

Bickler brought the motorized model rover over to Donna Shirley's
office to show it off. When he came into the front area, Shirley's secretary
took an interest. She thought it was cute! She wanted to know what Bickler called the rover, and how it worked. He didn't really have a name for it,
so he started to explain the running gear: "You've got the rocker here, see,
and then this front bogie . . ." "Rocker-bogie" became "Rocky."

✳

Bickler was proud of the rocker-bogie design, but he worried that it might
not be good enough. NASA had directed JPL to fund other organizations
in addition to itself through the Pathfinder Planetary Rover program.
Congressional mandate required that a large fraction of NASA research
funds be directed to private industry and universities, and the headquarters manager responsible for Pathfinder had specific ideas about where his
money should be spent. The consequence was that much more money
was funneled through JPL to other organizations than stayed at the Laboratory.

The primary beneficiary of these research dollars was Carnegie-
Mellon University's Field Robotics Center in Pittsburgh. When Bickler
thought about it, his expression was as if he had bitten into something
that tasted rotten. "The big money always went to CMU," said Bickler.

"JPL got a hundred thousand, and CMU got a million!" CMU's robotics research was led by William "Red" Whittaker, a hulking ex-Marine who was the research community's principal proponent of "big" robotics. The Pittsburgh team's forte was teleoperated dump trucks, earthmovers, and rugged vehicles for high-radiation or otherwise hazardous environments. Whittaker was a zealot. He wanted nothing less than the ubiquitous use of robots throughout human society, and would do anything to achieve that end. On first meeting him, I imagined that he must work twenty hours a day, and sleep only a couple of hours a night, an incongruous combination of exhaustion and energy, kept in balance by force of will.

Whittaker seemed to possess an army of "slave" laborers—graduate students and recently degreed engineers—ready to design and construct whatever concept Red proposed. For NASA they were building Ambler, a walking robot with six sixteen-foot telescoping legs. Bickler hated the design. "You don't really think you're going to send this thing to Mars!" But the advancing Ambler did make an imposing display.

Over the prior twenty years, walking vehicles had been created at several institutions, and some of them were impressive, going places wheeled vehicles could not. Mechanical complexity and low power efficiency were the bane of walkers. Electromechanical legs were inherently more complex than wheels, usually requiring several motors to the wheel's one. Walkers had to maintain their balance, and figure out where to place their feet in rough terrain. And because a walking robot tended to raise and lower its body with every step, it was as if it were always climbing stairs, which meant it was working harder than a rolling vehicle. CMU's Whittaker claimed that the Ambler design kept the main body at a constant height, so Ambler would "glide" over flat terrain efficiently. Bickler argued vehemently against this assertion. It was only after many months of development that Whittaker finally admitted the reality that each leg was sinking into the ground when Ambler shifted its weight to that leg; Ambler would indeed always be climbing uphill, burning energy.

Ambler faced the same weight problem as Robby, only more so. Carrying all of its computing and sensors onboard, it stood far above its human builders. It never did carry its own power supply, and even so weighed over six thousand pounds, more than a ton heavier than Robby.

One at a time, its legs would lift and swing around, then set down again, the body ever so slowly moving forward.

Bickler feared that JPL was going to lose its edge, that when the time came to build a real Mars rover, CMU would get the prize. He wouldn't mind so much if victory went to the institution with the best technical design, but it seemed to Bickler that "CMU was getting all the money to do stupid things!"

Robby and Ambler would eventually meet. They would face each other across a simulated Martian terrain during the NASA-sponsored Rover Expo in early September 1992, an event showcasing robotic rovers from around the world. The Rover Expo would be staged on the mall in Washington, D.C., across the street from the National Air and Space Museum. Robby and Ambler would be the giants of the show. By then, however, both would represent the past, not the future.

❊

Despite the superiority of Robby's hardware over the Blue Rover, Semi-Autonomous Navigation remained a slow, disturbingly time-intensive process. The bugs were being fixed. After integration and testing, Robby's hardware and software were now functioning as designed. Yet there were not enough daylight hours to traverse the required hundred meters (about a hundred and ten yards) in a single day. The best Robby could manage was thirteen yards in four hours of continuous operation!

Don Gennery's software was the culprit, requiring over an hour for stereo processing and terrain matching. Brian Wilcox tried to convince him to put effort into reducing the software processing time, but Gennery would have no part of it. "The first version of Robby was all Don's code. I couldn't light a fire under Don to speed up the code and to optimize it in any way. And it was causing us all kinds of political problems. People were saying we didn't know how to run projects, and everything else, which caused Don and I considerable . . . anguish." Gennery was an extremely able engineer, capable of producing major results in short periods of time. But he also suffered from "Ph.D. arrogance," devoting himself only to those activities that he himself deemed worthy of his talents. Gennery thought any reasonably intelligent person could recognize from the existing Robby capability that the key principles of SAN had been proven, and extrapolate the performance of Mars rovers assuming the faster computers that would obviously be available in future years, together with optimized code that other software engineers could generate.

Wilcox was still regularly having to explain Robby's rubber tires. You couldn't expect people to do much extrapolating. Appearance was all-important. In the minds of too many people, slow performance translated to "It doesn't work," regardless of the demonstration's technical merit.

We would have to do the demo again, and get it right this time.

Over the next several months, Robby's thought processes went through a readjustment. Larry Matthies, a Ph.D. in computer science from CMU, had recently been hired into Wilcox's group. He brought with him his own stereo vision algorithms and software, which he adapted for Robby. Unlike Gennery, Matthies was interested in speeding up his code. The terrain matcher module was dropped altogether.

When Robby went out into the field again, it was many times faster than before. It navigated the One-Hundred-Meter Milestone in a little over four hours.

✳

At JPL, Robby was the only game in town. Dave Miller wanted to change that. He had his own views about how rovers should work, views that did not jibe with Wilcox's sense-perceive-plan-act paradigm. But if he was to wrest control of rover research from the electronics and control division, Miller would have to start out small. Literally. At the Massachusetts Institute of Technology Artificial Intelligence Laboratory, Rodney Brooks was experimenting with tiny robots. Rather than try to duplicate human intelligence in any form, Brooks was attempting to mimic the behavior of much more modest creatures. Insect brains had relatively few neurons, yet insect life swarmed over the planet, among the most successful creatures on Earth.

Brooks's concept was that one could layer a number of extremely simple behaviors, one atop another, and produce useful results. Miller didn't want to shrink rovers down to Brooks's realm of robotic insects. Shoebox size would do. Miller's group built a four-wheeled tabletop rover they named "Tooth." It would never survive outdoors, but it could demonstrate a few behaviors, responding to signals from simple photo-detectors and contact switches.

Miller was desperate for a terrain-capable rover. Thanks to Bickler's mechanical team, there were now ready-made vehicle suspensions just waiting to be used. Miller and John Loch, an engineer in his group, first added a simple computer to the Rocky vehicle to control its wheel drives and steering motors. They could then joystick Rocky around, driving over rocks and impressing onlookers.

Bickler's experimentation with Rocky had led to an improved design called "Rocky 2." But that design had been scrapped before it was ever built. Rocky 3 had come off the assembly line with bigger wheels that wouldn't sink so easily in soft sand, and proportions optimized for climbing over rocks. Other people might look at Rocky 3 as a model of something bigger; to Miller, the vehicle was already full-scale, a "microrover" ready to compete with Robby.

Miller needed to convince the right people of the flaws in the Robby "big rover" approach, and of the advantages of the microrover. So he lobbied the JPL manager of Pathfinder Planetary Rover. He lobbied the NASA headquarters sponsor of the activity, and he lobbied Donna Shirley. Miller carefully promoted the revolutionary nature of the microrover: It would be cheap, cute, sexy, and imbued with AI magic, in the form of what he called "Behavior Control." Megarovers like Robby were slow, lumbering, and overcomplicated.

Ironically, the work of Rodney Brooks that had inspired Miller to the idea of microrovers had itself been partially triggered by Brian Wilcox a few years before. Wilcox had presented a "Micro-Lunar-Rover Challenge" at a NATO robotics workshop in Portugal in May 1987, several months before Miller joined JPL. Wilcox had proposed the idea of sending rovers massing less than twenty-five pounds to the Moon, at a time when future Mars rovers were expected to be over two thousand pounds. Brooks had taken the ideas home from the conference, while Wilcox had returned to JPL to support the ongoing MRSR rover studies.

✳

At the direction of the NASA sponsor of Pathfinder Planetary Rover, a second prong was added to the research program's focus: Demonstrate a microrover that returns a sample to a simulated lander. Miller's Robotic Intelligence group already had a simple computer on wheels, using Bickler's Rocky 3 vehicle. Rajiv Desai built a crude arm with a sampling scoop that the team mounted on the front of Rocky 3. Their team integrated the hardware and software into a new rover system.

When all was ready, the rover autonomously dead reckoned to a specified target, using a compass and measuring wheel revolutions to estimate its position. Once the onboard computer had estimated that the rover had reached the sampling site, it halted the traverse and dropped the end of the arm onto the surface. To ensure that the gripper had fallen on soft soil rather than on a rock (which would probably not be retrievable) the scoop contained a pin sensor. The pin was designed to activate a micro-switch. If the pin struck a hard surface, like a rock, the switch would be triggered, and the rover would lift the arm and drop it at a possibly more favorable spot. If, instead, the pin had entered soft soil, the switch would not be activated, and the site would be deemed "good." The scoop closed, the arm was raised, and the rover traversed back to the lander, homing in on an infrared beacon. When it got close enough, the rover dumped the contents of its scoop into a cardboard box that was a stand-in for a lander's sample collection bin.

✳

The MRSR mission study had become a means for each research discipline to justify its latest research efforts, and market its NASA sponsors for increased funding. The rover, carrying radioisotope thermoelectric generators for power, multiple robot arms, and a sample-processing system, had grown to 1,100 pounds. It would cover great distances, and in the end collect up to ten pounds of rock samples for return to Earth.

The price for those rocks would be high: Our best guess was $10 billion. An outside contractor was hired to produce an independent cost estimate of the JPL concept. Their numbers were worse, anywhere from

$10 to $13 billion. That was big, five times bigger than any robotic deep space mission we'd ever done. And the support for MRSR started to evaporate.

Too late, we realized the day of the big mission was over. MRSR was the dinosaur that could not adapt to the new environment after that asteroid punched through the atmosphere into the ocean. The weather got colder, the skies clouded over, and the advantage went to the small.

※

SIX

THE LITTLE ROVER THAT COULD

Mars Rover Sample Return had collapsed under its own weight. This left a huge gap in NASA's plans for the further exploration of Mars. Within a few months, major boosters of the MRSR mission were disassociating themselves from the study and the politically unpalatable price tag it had generated for the mission. Was surface exploration of Mars dead?

Another NASA facility—Ames Research Center—stepped into the breach with a proposal for an entirely new series of missions, dubbed the "Mars Environmental SURvey," or MESUR. MESUR would be a set of sixteen to twenty landers that would form a network of science stations blanketing the entire planetary surface. The landers would collect seismic and weather information, enabling scientists to construct a global model of Mars. The cornerstone of the mission set would be a small lightweight landed station which would be replicated many times. The landers would be so small in volume and mass that, at least as proposed, four could be sent to Mars at once on a single, relatively low-cost Delta II launch vehicle. Given this approach, Ames estimated that all of MESUR could be accomplished for less than $1 billion. This was an impressively modest price tag after the sticker shock of MRSR.

The proposed MESUR might not be a JPL mission, but it could yet prove to be an opportunity for JPL technologies. JPL would show the Mars science community and NASA headquarters that the Lab was the place that could deliver the best, most exciting surface science. The head of the Office of Space Science and Instruments at JPL brought together three key people at JPL for weekly lunch discussions. One of them was a micro-devices technical guru, who was using new fabrication techniques to create incredibly tiny instrument packages. Matt Golombek was one of the Lab's leading planetary geologists. And Dave Miller was carrying the banner of microrovers. The small lightweight MESUR lander would have room for only an equally diminutive payload. To capture the backing of the scientists, that payload would have to be capable of performing exciting science. These conversations, which soon included Brian Wilcox, led to a concept for constructing and operating such a payload, not just promising it on paper. They would build a rover massing only a few pounds, and the instrumentation it would carry: a camera, point spectrometer, rock chipper, soil scoop, and a micro-seismometer only a few inches on a side. The rover, operated from a simulated Earth station, would perform a complete science mission. It would emplace the micro-seismometer on the surface, then traverse to a rock. The chipper would wear away the rock's outside surface, revealing the pristine material underneath (which was of much greater interest to geologists than the outer "weathering rind"). After taking spectrometer readings of the rock to assess its composition, the rover would scoop up a small soil sample and deliver the sample back to the mock lander from which it had originally descended. Funds from a variety of sources would be pooled, including JPL's scarce discretionary moneys, existing rover research dollars, and an additional infusion from Headquarters, together eventually comprising the approximately $1 million necessary to demonstrate a new rover capability. It would be called the Mars Science Microrover.

Initially, Golombek was given an allocation of about $300,000 to get the effort started. At first, the technical approach was unclear: Wilcox and Miller were both given access to the funding, but little direction.

Golombek was a scientist, not a task manager. With no one supplying the technical leadership from the top, Wilcox fell into a role he had taken on several times over his career: providing leadership from below. He chose a technical approach and proceeded to implement it in a quick, cheap, and this time, not so dirty way. Perhaps others would follow his lead, perhaps not. But MSM had no chance of having anything to show by the next summer unless they got moving soon. Wilcox's plan also had the benefit of giving his Robotics group a microrover of its own to compete with Miller's adaptation of Rocky 3.

To keep the weight of the rover down, Wilcox figured you could leave the computer off the rover entirely. "It didn't really matter where you put the computer, so you might as well put it on the lander," at least for this demonstration. The lander was going to have its own computer onboard anyway, so the rover could take advantage of it. The rover would then be mostly a radio-controlled car operated by the lander, with a few more actuators thrown in to operate a sample scoop, and stereo TV cameras to send back pictures.

One of the bugaboos of off-roading robots was the danger of flipping over. While a rover could be outfitted with tilt sensors to warn itself of an impending tip-over, you could never guarantee the vehicle wouldn't slide off a particular rock and end up on its back. Whenever JPL opened its gates to the public for an open house, someone looking over the rovers on display would inevitably ask, "What happens when that thing rolls over?" The ideal rover would be "self-righting," somehow putting itself back on its wheels to get out of a jam. Wilcox had a concept in mind for a four-wheeled rover that wouldn't care whether it ended up upside down, and might be able to handle terrain as well or better than Bickler's six-wheeled Rocky series. The rover looked like it was rolling around on stilts. Working together with Timothy Ohm, a gifted mechanical engineer who was also an expert machinist, Wilcox designed and built "Go-For," a fork-wheeled mobility system. The front "stilts" were both mounted to the same axle, which had its own motor drive and could be rotated through a full 360 degrees. The rear stilts were similarly locked together, creating a "fork" with wheels on the ends of the prongs. The result was that the vehicle could actually do slow somersaults, turning itself over when necessary. And by shifting the forks through smaller angles, Go-For's weight could be transferred from its front to its rear wheels. To go over a rock, Go-For could lean back, taking the weight off the front wheels, making it easier to lift the front onto the rock, while simultaneously putting more weight on the rear wheels, giving them more traction. Once the front wheels had a good purchase again, Go-For could lean forward, now making it easier for the rear wheels to climb the obstacle.

Miller was extremely unhappy with the mere existence of Go-For. While Wilcox was marketing Go-For by carrying it to offices and driving it around, Miller lobbied against its use on Mars Science Microrover.

One day Golombek instructed Wilcox to stop charging to the MSM account. Golombek told Wilcox that a new task manager, Arthur "Lonne" Lane, was about to take the reins of MSM; Lane would decide what the proper next steps would be.

Lonne Lane and his skunkworks-style team had just come off of a task called Delta Star, which had produced flight hardware from start to finish in only fourteen weeks. Lane seemed to be a good choice to pull together the disparate elements of a complex rover system on a tight schedule. After all, the MSM rover did not need to meet the standards of flight hardware; it only had to appear plausibly flightlike to the appropriate audience, and function properly on Earth. As a goal, Lane selected a total mass of eighteen pounds for the MSM rover. The decision was not arbitrary: The MESUR lander could probably afford to carry a payload of this mass. If Lane succeeded in shrinking the rover as planned, Rocky 3 (which had grown to forty-four pounds) would look like a clunky dinosaur in comparison.

In September 1991, when Lonne Lane began leading the MSM effort, he had no experience whatsoever with rovers. He had not worked with any of the rover engineers who would be key to building the demonstration system. Lane did bring with him a rarity of experience: He was a scientist who had delivered flight hardware. He understood the relevant science community, and knew how to run a flight project development team, neither of which the rover researchers knew much about.

Lane nixed Wilcox's Go-For concept. It turned out that it did matter where the computer was, at least when you had sophisticated science instruments on the rover. You needed the computer to operate the instruments and process their data. Go-For had the disadvantage that to get the most out of its mobility capability, you needed some kind of active control that constantly monitored where the weight of the vehicle was shifting and modified the fork angles accordingly. Rocker-bogie vehicles were generally more stable, with no added onboard "intelligence" required.

Lane got a crash course in rover mechanics from Don Bickler. A group of mechanical engineers was tasked with creating the next in the series of rocker-bogie vehicles. The Mars Science Microrover, alias Rocky 4, would have about the same wheelbase as Rocky 3, but would weigh a third as much. A common method to get weight down in structural members is to make them hollow—a hollow rectangular beam can be almost as strong in most cases as its solid counterpart, while containing only a fraction of the mass. But manufacturing hollow structures of unusual shape can be costly and difficult, and MSM money was tight. Faced with the challenge of keeping both weight and cost low, Bickler would end up inventing a new machining technique.

In the competition for who would build the rover "smarts," Dave Miller's group got the gold ring. Wilcox's group would create the control station and the simulated lander. The decision was presented as a simple division of labor, making the best use of the cadre of rover engineers. My reaction, however, was one of dismay. Our group had been building and controlling rovers for years, and knew the relevant technical issues intimately, but now we had been relegated to a support role. Rocky 4 would be directed by Miller's 'behavior control.' Would we be shut out from now on? Wilcox advised me not to worry: We had gotten the better, more challenging assignment. The Robotic Vehicles group would still be responsible for commanding the rover, even if the Robotic Intelligence group would build the onboard control system.

Once again, Brian Cooper would be responsible for designing a vehicle control station. Like the Army-funded Robotic Technology Test Vehicle, and the Blue Rover before it, MSM called for a version of CARD. But this time there would be a new challenge. In each of the previous implementations, the stereo cameras had been mounted on the vehicle, and moved with the rover. But the MSM rover would be only about a foot tall.

The single camera onboard would be so low to the ground that Cooper would hardly be able to see any distance ahead to plan a path. A rock just a few inches tall might totally block the view. The MESUR lander would almost certainly have a camera anyway, and it would be sitting at the top of a tall mast, with a much better vantage point than the rover. When the cameras were fixed to the vehicle, you always commanded relative to the current position, such as "Go to a spot one yard forward and three yards to the left." If you used the lander's cameras, you saw more, but now the rover was always moving relative to the cameras' location. Not only would Cooper have to tell the MSM rover where to go, he would have to tell it where it was starting from. And since the cameras stayed behind with the lander, it would get harder and harder to accurately designate the rover's position and target locations as it got farther away. The microrover also had instruments onboard that would need to be commanded. The new control station would issue commands to the spectrometer to gather data, to the rock chipper to start and stop chipping, and to the scoop to pick up soil.

Others in the Robotic Vehicles group, including myself, designed and built the MSM simulated lander. We called it "simulated" because it didn't actually land; it was a platform that performed only the functions a real lander would do to support the microrover operating nearby. Except for the lack of wheels, the MSM lander was similar in concept to Robby. A pair of small stereo cameras was mounted on a compact pan-tilt mechanism elegantly designed and machined by Timothy Ohm. Electronics in the guts of the lander could capture frozen images from those cameras and transmit them back to Cooper's control station. A radio modem located near the cameras would transmit commands to the rover, and receive data sent from the rover. Ramps located front and rear gave the rover a way off, and back onto, the lander, while a trough in the middle provided a spot for Rocky 4 to deposit collected soil samples. When the microseismometer arrived, it would be tethered to the lander's electronics, which would communicate the readings to a science display station. We covered the lander's external panels with gold Mylar foil bought from the local sporting goods store for an appropriately spacecraft-like appearance.

Lane aimed the team at a mid-summer demonstration. To achieve it, each subsystem would have to deliver its part on time. The new micro-rover chassis, computer, and electronics were due at the end of March 1992. Allowing until the end of May for installing and integrating the elements together, there would still be two months for finalizing the software and testing with the lander and control station. The schedule looked . . . possible. Would we be able to pull it off?

✳

At the end of October 1991, Matt Golombek and Dave Miller flew to Washington, D.C., to pitch the Mars Science Microrover at the Sixth Mars Science Working Group meeting. This group was the team of planetary scientists that advised NASA on science priorities for future Mars missions. From the science advocacy standpoint, these were the very people Mars Science Microrover had been designed to impress.

Much of the two-day meeting focused on the MESUR mission. The science community was concerned: Would MESUR do credible science? For engineering reasons, sampling of subsurface materials was not part of the mission designers' plan. That left dust, soil, and surface rock. A potentially important instrument for MESUR was the Alpha Proton X-ray Spectrometer, which was capable of determining the elemental composition of rocks and soil, answering the fundamental question, "What are they made of?" To give meaningful results, the APXS would have to be placed in direct contact with the target to be analyzed. But how do you get the APXS onto rocks? And since you would have brought the instrument all the way from Earth, you really wanted to put it on more than one rock. The chief engineer for the MESUR study at Ames suggested several options for deploying the APXS from the lander, which included dropping the APXS on the surface, catapulting the instrument out from the lander, catapulting it and then reeling it back in, aiming the catapult first, and finally the possibility of placing the instrument about three feet from the lander using a robotic arm. The Working Group scientists viewed these ideas dimly. They were so unimpressed with the options that they agreed to consider replacing the APXS with other instruments that did not require placement in such close proximity to their targets.

Scott Hubbard from NASA Ames presented the concept of SLIM

(Surface Lander Investigation of Mars) to the working group. This was a proposal for a "mini-MESUR," a single lander that would be launched in 1996, four or more years sooner than the first proposed MESUR launch. The hockey-puck-shaped lander could land either right side up or upside down: The instruments would be configured to deploy properly either way. The design (and the science) would be simplified to bring the cost within the $150 million limit imposed for NASA's new Discovery-class low-cost missions.

This was the first time anyone in the Science Working Group had heard of SLIM. Certainly nobody at JPL was aware of it prior to the meeting. Although SLIM would put a lander on Mars many years earlier than the MESUR mission, the working group wondered, what good was a single lander? The concept was not received well.

Golombek's microrover presentation was not scheduled until the second day of the meeting. He was nervous, and could not sleep the night before. This would be one of his first presentations in front of a large group of science heavyweights. "I'm going to get toasted alive," was the thought in his head. At 3 A.M., Golombek found himself hand-lettering a new introductory viewgraph for the start of his talk. He knew that he had to break through the perceptions that the MRSR rover study had created within the science community. The MRSR legacy was that rovers needed to be large and complex to be useful; they were difficult to operate, requiring a major infrastructure; and they were expensive.

Morning came. At his presentation, Golombek addressed the negative perceptions head on. His premise was that the belief that surface mobility required a rover the size and complexity of MRSR was just plain wrong. Rovers could be small and simple, easy to operate, and cheap. The APXS just cried out for a microrover to deliver it to rocks. And JPL was going to prove that a small rover could do useful science within another nine months. Mars Science Microrover was already under way.

Golombek listed the existing instrument payload for MESUR, and proceeded to describe what a microrover could do for that payload, improving the science you could get back. A rover could get up close to Mars rocks, providing the equivalent of a geologist's hand lens. It could inspect many rocks, not just one. Rockets firing during landing tended to create a contamination zone immediately around the lander. A microrover could traverse outside of this zone, retrieve soil or dust, and bring it back to the analyzers on the lander. MESUR called for the placement of many seismometers, one per lander. Vibrations from the lander itself or the Martian wind could corrupt the seismic measurements; the rover could emplace a small seismometer safely away from the lander and potentially out of the wind. And finally, a rover would allow the science team to actively explore the landing site, going out to features that looked interesting, instead of leaving them tantalizingly out of reach.

Dave Miller followed Golombek with a talk about the feasibility of small rovers. He showed a videotape that demonstrated Tooth, the four-wheeled tabletop robot about twelve inches long. Tooth looked like a toy, but it actually could perform a crude mini-mission. Photodetectors onboard could determine the direction of a bright light, so Tooth would drive away from such a light. Tooth also had a very simple gripper on the front, one that would close if any object blocked the light source on one gripper jaw from the light detector on the other. You could stick your finger between the gripper jaws and they would close on the finger. But once the gripper fully closed, it would reject the finger as too small and let go. Tooth was after bigger prey: the plastic cap to a spray-paint can. When the gripper could only close partway, Tooth knew it had what it wanted in its grasp. So Tooth would wander away from a light until it ran into an appropriately sized plastic cap. Once it had grasped its "sample," its behavior changed: Now it was enamored of light, and it headed back home toward the light source. When the light source got bright enough, Tooth changed its mind again, deciding it didn't want the cap anymore. The little robot set down the cap by the light, then wandered away, seeking more caps. Given enough time and caps in the vicinity, Tooth could collect any number of caps and deposit them all near the light. If you imagined the light source as a point on the lander, and the caps as small rocks, the relevance to MESUR seemed clear. Miller had brought Tooth with him, and he demonstrated some of Tooth's simple behaviors, including its proclivity to grab hold of his finger and then politely let go.

By the end of the meeting, the Science Working Group's recommendation to NASA was to add a microrover to SLIM, and go for a launch in

1996 if the funds could be found for a "quick Mars mission." The rover would make a single lander like SLIM useful, and provide a way to deliver the APXS and potentially other instruments to their science targets.

On the way home, Golombek was almost literally jumping up and down. Before the Working Group meeting, JPL had been pretty much cut out of future Mars exploration. Now it had the hot technology that made a '96 Mars mission worth doing.

Within a year of Ames's original MESUR proposal, and only two months after the Mars Science Working Group meeting, JPL became the NASA center for the new Mars Exploration Program, and the place where MESUR would be implemented. While the mission had still not been approved, NASA had handed the charter to do it, if and when the mission became real, to JPL.

The date of the Mars Science Microrover demonstration had only been vaguely defined as middle to late summer. The objective had been to develop a compelling microrover capability in time to influence the funding cycle for the next fiscal year. Then, near the end of April, in the midst of dealing with delays in the completion of the Rocky 4 mobility chassis, Lane received a new directive: the MSM rover was to be demonstrated as the centerpiece of the celebration of the twenty-fifth anniversary of the first Surveyor soft-landing on the surface of the Moon, to be held on Friday, June 26, 1992. That was barely two months away!

There was still no rover body to install the computer and instruments onto. The MSM group, already working hard, was pushed into a frantic mode. Lane began holding meetings every morning. The group would assemble in Building 107, standing around while Lane or the system engineer outlined the schedule for that day. The mechanical team finally delivered the rover chassis with flightlike stainless steel wheels in May, over a month after originally promised. Could the Robotic Intelligence group get its computer and software installed and working in the six weeks remaining?

The contracted-out computer electronics boards did not make their job any easier. To fit into the limited volume available on the Rocky 4 chassis, the rover brain had been envisioned as a cube a few inches on a side. The walls of the cube were prototype boards with the necessary electronic components mounted to them. There wasn't enough time to manufacture proper circuit boards, so the circuits would have to be "wire-wrapped" with individual wires leading from each component to the next. With hundreds of wires crisscrossing over the boards, the computer cube would look like a bird's nest. Lane didn't feel that Miller's group had the time to build the computer themselves, so to do the job he had selected a local company often used by JPL. They usually did excellent work, but when the cube arrived, it was a mess. The wires had been wrapped too tightly and tended to break off of the component pins at the point where the exposed wire met its insulation. Worse, although wire-wrap wire was available in many colors so you could tell one from the other in just this type of situation, the contractor had used only one color. Debugging the boards would be nearly impossible! But they did it anyway.

As the hardware and software integration raced forward, the logistics of the demonstration began to come together. Since the rover was at the core of the festivities, Lane's MSM team had been handed control over the orchestration of the event. Permits were sought. Fences went up around the site. (Although both rovers and rockets had been tested in the vicinity, the Arroyo Seco itself was not on JPL property, and in fact belonged to the city of Pasadena.) To acquire "descent imagery," showing the equivalent of what a Mars lander might see on its way down, Ken Manatt, one of the engineers on the team, flew his hang glider from the top of neighboring Mount Wilson to JPL, snapping pictures as he maneuvered over the rover operations area, and finally landing in the debris basin just south of the Laboratory.

Before the team was ready, Lane forced the group to move its operations to the outdoor demonstration site. He knew that the system had no chance of working unless the team was familiar and comfortable operating outside well before the big day.

*

The Robotic Vehicles group and the Robotic Intelligence group had difficulty collaborating. From the days of Robby, the AI team viewed the rover navigation testbed as a behemoth, and the approach of modeling the terrain around the vehicle in detail as misguided. When talking about the Robotic Vehicles group, the common refrain within the AI team was an exasperated "They don't get it!" One by one, their arrogant attitude alienated the other members of the MSM team. Now that the AI team was in the crunch to deliver a working rover, they were not so confident. This was their opportunity to demonstrate the reality of their vaunted "behavior control," but it was also a chance, independent of the veracity of their ideas, for them to fail. Some things that should have worked didn't, and some things they just didn't have time to do in the first place. Readings from simple sensors were supposed to trigger behaviors that led to a safe traverse that avoided rock hazards. They had mounted small sensors on the corners of the rover that radiated infrared light and looked for some of that light to be bounced back. If a sensor detected some reflected infrared, this would indicate that Rocky 4 was approaching a rock hazard and needed to veer off. But the infrared light detectors seemed to be blind to many rocks. There was just not enough light being reflected back. If the rocks were effectively invisible, there would be no sensor readings to trigger a new behavior, and Rocky 4 could bump into things. There was no time to come up with a new sensor type, so the only strategy left was to move enough rocks out of the way to leave a clear path for the rover. The Robotic Intelligence group was under extreme pressure to perform, and would sometimes respond to that pressure by blaming their woes on others, legitimately or otherwise.

The demonstration site was located in the Arroyo Seco just east of the Laboratory. The MSM team created a miniature rock field by carefully arranging rocks of several sizes brought in for the purpose. Large open-sided tents were erected nearby so that the VIP audience and media would be able to sit comfortably out of the summer sun. The rover control center was located within a large rental truck, with shock-mounted computer workstations and video monitors strapped to the inside walls. During testing we would often leave the rear roll-up door of the truck open, but during the actual demonstration the door would be shut, forcing the operators to rely only on lander camera images and video from the rover-mounted camera. In addition to the rover control computer, another workstation was dedicated to processing and displaying the science data from the micro-seismometer and the rover's onboard spectrometer. Cables ran from the computers inside the truck to video monitors in the audience viewing area. Outside, the simulated lander had been precisely situated at one end of the rock field, its ramps leading down into the constructed terrain.

Soon, another lander stood nearby: JPL's full-sized model of Surveyor had been carefully trucked out to the Arroyo site and situated just outside the rover's test course.

On the day of the demonstration, I found myself sitting in front of the rover control station. Months before, Brian Cooper had planned an expensive vacation with his wife, before the demonstration date solidified at just the wrong time. Cooper had nonrefundable airline tickets. The vacation won . . . and now I was in the hot seat. Brian Wilcox sat behind, the backseat driver who would have time to think while I was busy typing commands.

In the week or so before, Rocky 4 had successfully performed all of the steps of the demonstration. To get more test time, we had set up floodlights to illuminate the test course and often operated late into the night. The trickiest part was getting the microrover positioned just so in front of a target rock. Every time you directed the rover to drive some distance to a desired spot, slippage of the wheels in the soil or on rocks caused it to end up in a slightly different place. Looking at the video from the rover's own camera, you could estimate how much you'd have to turn the rover to face its target, and then move it forward to put the chipper against the rock surface.

Our usually deserted test area was now bustling. JPL and NASA officials had been bussed out to the site. Television crews and press reporters had come with them.

The crowd followed Lonne Lane over toward the simulated lander. Rocky 4 had been placed on the top of the lander, facing backwards. Its rock chipper was a robotically operated and aimed ice pick sticking out the front of the vehicle. The chipper was now tilted upward, as if to warn away the inquisitive giants that might threaten the little rover. Dangling from a hook slipped over the chipper was the tiny but functioning microseismometer.

By command from inside the truck, I turned on the seismometer and the rover's camera. By glancing over at the video monitors, the audience could get a rover's-eye view of the events to come. Certainly Lane was explaining this and much more to the assemblage, but I was oblivious to everything except the signal to start the demo. The signal came and off we went. I sent the "Rover Disembark" command, which was relayed from the control station to the lander, and thence over the radio link to Rocky 4. And off the rover went, rolling rapidly down the ramp onto the sand, the seismometer held high and safely out of the way. The seismometer trailed its data transmission cable behind it. Rocky 4's backwards driving avoided trampling and potentially entangling the cable. Next Rocky 4 lowered the chipper until the seismometer rested in the dirt, then rolled backwards another eighteen inches, leaving the seismometer deployed where it belonged.

So far so good. We then sent the rover on to two designated waypoints that would get it to the rock of choice. We got Rocky 4 aimed at the rock, drove it forward, and tilted the camera-spectrometer platform up for a good close-up of the target. Spectrometer data flowed in. Now it was time to activate the chipper, which would batter its tip against the rock, removing the surface layer in preparation for getting another spectrum.

The rock chipper started up. It rattled away at the rock. And it just kept going. No message came back from the rover to confirm the command. I scrambled to send an "Abort" command. No effect: The rover was merrily chipping away. The only thing left to try was the "Emergency Shutdown" command. I knew if I sent the shutdown, the demonstration was over: It would take too long to re-initialize the rover after that. Wilcox and I conferred for a few seconds. "Send it." Nothing happened. Rocky 4 was no longer responding to commands. The rover's computer had crashed, and there was nothing more we could do. The chipper ran on, oblivious.

The audience began to realize that the demonstration was not proceeding properly. After a moment, Lane strode over to the rover. He reached down and shut off the rover power. The embarrassingly persistent chattering of the rock chipper stopped. Lane calmly explained that the rover was having a problem, quite possibly due to radio frequency interference from all of the news crews' video cameras nearby.

Inside the truck, I was stunned by the failure. We had never had such a problem during testing, and at just the worst moment the rover had died. The press would have a field day!

Outside, Lane continued to tell the story of the rover. He asked the audience to imagine the next steps the rover would follow, extrapolating from what they had already seen. Although temporarily halted in its mission, the microrover had already done real science. On the video screens in the tent, you could see the images the rover had taken with its onboard camera. There was the first data set from the mini-spectrometer. And the micro-seismometer deployed by the rover was generating results even now. To accomplish its full mission, the rover need only traverse to the nearby sandy area, scoop up some soil, and return it to the lander for analysis. The JPL Director, the NASA Associate Administrator, and the reporters saw the little rover as a success. Lane knew how to talk to his audience and give them a piece of his vision. He later commented, "It helped that no one had any expectations of a microrover's capabilities to compare with."

Rajiv Desai, Dave Miller, and their team were livid in the face of their perceived fiasco. Once the crowd had moved on, they took Rocky 4 back to the staging area and attempted, unsuccessfully, to revive the vehicle. They formulated a hypothesis as to what had probably happened: The vibration of the rock chipper had caused a short in the poorly wired computer board, frying the CPU chip. The CPU had turned on the chipper just fine, but there was no brain left when the time came to shut it off. Rocky 4 had effectively committed suicide. In frustration, one of their team pulled the CPU chip out of its socket on the board, threw it to the ground, and possibly stomped on it. Desai and Miller blamed the hardware, and Lane's poor choice of contractor to build it.

Ken Manatt picked the chip out of the dirt. He took it back to the lab for a more thorough evaluation. When he powered it up in a benchtop test environment, the chip at first would not operate. The small onboard memory, which told the chip how to initialize when it was first turned on, seemed to be corrupted. Once Manatt reset the memory to its factory settings, the chip functioned normally, despite the mistreatment it had received. Perhaps there had been a short in the wiring of Rocky 4's computer boards, but it had not been sufficient to permanently damage the CPU.

Lonne Lane's own surmise was that the rover software team had simply not been given enough time to integrate and test the low-level software that drove the various onboard devices, and the computer had simply crashed. In his view, it might not have been humanly possible for anyone to have fully completed the software job, given the few weeks available between the late arrival of the rover chassis and the immutable date of the demonstration.

The definitive cause of Rocky 4's seizure would remain a mystery. The sweep of events to come would leave little time for a detailed investigation, nor a strong need. For Mars Science Microrover had delivered on JPL's investment. With a margin of seconds, Rocky 4 had survived just long enough. Despite the early termination of the demonstration, MSM had proven the viability of the microrover as a component of Mars surface exploration.

A few days later, Donna Shirley called an all-hands meeting for the MSM team. First she congratulated the team for a job well done. Then she moved on to the real purpose of the meeting.

There was going to be a new mission to Mars. It was called MESUR, for Mars Environmental SURvey. It was really a series of missions that would put many landers on the Martian surface. The first launch would demonstrate the technologies that would be necessary to make the entire mission set work; it was sort of a trailblazer for MESUR, so it had been named MESUR Pathfinder. (The project had no connection with the planetary rover research program that had produced Robby. "Pathfinder" just seemed to be a popular name in NASA circles.) The mission had not yet been approved, but that would come soon enough. Meanwhile Code R, the NASA organization that among other things controlled NASA's funding of automation and robotics research, was going to pay for the development of a *flight* microrover similar to the Mars Science Microrover, intended to fly on MESUR Pathfinder. This was unprecedented: Code R did not fund flight systems.

Of course, sending a rover to Mars was also unprecedented.

Shirley announced that she was to be the rover project manager. Over the next few months she would be assembling the flight team, and many of those in the room today would participate.

Code R was going to "pony up" $25 million for the rover. That was a huge amount for a research task, but it was not very much for a flight system. The new flight rover team would be lean. It needed to get started now, to get a jump on the MESUR team. When MESUR did get its act together, and was officially approved, it would be a significantly larger activity, moving much faster than the smaller microrover team could ever hope to do. The flight rover team would need a head start. The rover effort would have to build up its own momentum soon, so that while MESUR would inevitably catch up with it, the rover would not be left behind. In fact, the people planning MESUR had not yet even promised to fly the rover onboard their lander. But the lander had to carry some kind of payload, and the rover would be ready.

I had always wanted to be part of a mission that explored space, like the Mariner, Viking, and Voyager missions I had watched and read about while growing up. I had held that secret hope even as I first came to work at JPL. But over the years of working on rovers, I had come to accept that a real mission would never happen. Each research task taught us more about how to make rovers work, but seemed to lead only to more research.

Only at this moment, as Shirley described the plan, did the idea of a flight rover become real to me. There were still doubts: Missions could get canceled—or never approved in the first place—and no one had told me personally that I would be part of the flight team. But there were plans, deadlines, and money, all aimed at building a new microrover, and the big word "FLIGHT" was stamped all over them.

PART 2

PATHFINDER

SEVEN

A SMALL ENOUGH TEAM
TO DO THE JOB

To begin work on a flight project is to enter a new world where mass, power, and volume are precious commodities. Consuming too much of any of these is not an option. Each available rocket—in aerospace they call them "launch vehicles"—whether a Delta, Titan, or Ariane, has only so much weight of payload it can put into a particular trajectory in space. If you are launching a spacecraft to Mars, and its mass is too high, the laws of physics ensure that it will never reach its target. Each spacecraft must carry with it its own power source, whether in the form of solar arrays, radioisotope thermoelectric generators, or batteries. These power sources are limited in their capacity. The components of the spacecraft that depend on this power must use it efficiently, for when the needs of the system exceed the available power, the spacecraft dies. And each spacecraft has to be small enough to fit within the launch vehicle's fairing, the nose section that protects the spacecraft and reduces aerodynamic drag during the rocket's supersonic flight up through the Earth's atmosphere.

None of these commodities had been critical in the rover research tasks, where the usual final product was a videotaped demonstration. And while the Mars Science Microrover had been designed to appear flightlike, that vehicle only had to operate for about an hour under its own power while surviving the very Earthly environment of the Arroyo Seco.

The flight microrover would be different. The constraints imposed on the Pathfinder spacecraft were trickling down to each payload element. The flight rover would have to fit within the tight confines of the Pathfinder lander, survive the rough trip to Mars, and operate on a planet where the surface temperatures ranged from a high of about 60°F to a low of −130°F. Somehow the rover would have to supply itself with power to run its computer and motors for days or weeks, not merely hours. The rover would have to do more than merely survive: It would carry an as yet unknown set of science instruments, and perform an as yet unknown mission. The whole package must fit into whatever mass allocation Pathfinder could spare. This would probably amount to some ten or so pounds.

It would be a lot to stuff into a small box.

✳

Donna Shirley was a manager with no one to manage. Only a few of the engineers on the former Mars Science Microrover team had the proper skills for the flight effort. Through her own efforts Shirley had achieved the position of Rover Team Leader, but building the rover team would not be as simple as calling up the people she had in mind and putting them to work. First she needed to identify who the right engineers were and negotiate their availability, convincing them to leave their current JPL jobs to join the rover team.

Little money would be available in the first year. But in that time Shirley's team would plan the schedules, determine the budgets, and make the key design choices to prove the feasibility of delivering a flight rover on-time and within the $25 million cost cap. The rate of spending—the "burn rate"—would jump in future years as the detailed design and implementation phases began; but there would be no more money after the $25 million was gone. This total cost was NASA's number one condition on the rover: If Shirley's team came up with a design that exceeded it, then they would go back and simplify the rover, making it dumber and slower if necessary, until they had an acceptable sticker price.

Shirley wanted lead engineers for each of the rover subsystems: Power, Telecommunications, Mobility-Thermal-Mechanical, Control and Navigation. She needed good people and she needed them right away.

She went to Charles Weisbin, manager of the Automation and Control section. Shirley wanted one "Cognizant Engineer" to be responsible for all of the control and navigation area. This subsystem would provide the rover brain, the software that would run on it, any sensors the rover would rely on, and the ground control station that would tell the rover what to do. Did Weisbin have anybody in mind?

There were no rover control engineers with previous flight experience—there had been no prior robotic flight rovers—so the question became which of the robotics research engineers in the section would be best suited to move over to the flight side of the house. Weisbin selected Henry Stone, another member of Wilcox's Robotic Vehicles group. Stone was then the manager for the Hazbot task, in which he had proven his mettle by adapting a commercial robot platform to operate in hazardous environments on Earth. He had also shown a knack for digging deeply into the details of whatever project he focused on. Weisbin thought Stone could well apply those traits to the flight rover. When Weisbin asked, Stone jumped at the opportunity.

Stone was impressed with how both Donna Shirley and Pathfinder "were putting together the project in a completely different way. There were no rules, or if there were, they were to be broken. It was just a gung ho bunch of rebels out there who were going to make that thing happen." He wanted to be a part of it.

Soon Stone was hunting for the system engineer for his new Control and Navigation subsystem. He did not know me well, but after some consultation with Brian Wilcox, he came to my door. Henry didn't have to ask me twice. I had my place on a Mars rover flight project.

✳

Howard Eisen joined the team as Stone's counterpart on the mechanical side. Since his days as a co-op student experimenting with Robby and Robby Jr., and working with Bickler on the development of Rocky, Eisen had completed his degree at M.I.T. and returned to the JPL Mechanical

Systems division as a permanent employee. He had already participated in two JPL flight projects for Earth-orbiting satellites. He was young, still only in his mid-twenties, but he had already shown himself to be a gifted mechanical engineer with a knack for getting things done. He could be brash and abrasive, but he made up for it with creative engineering skills and an intuitive grasp of the relevant technical details. He and Shirley got along famously.

※

Shirley continued building her team. Upper management told her the name of one more person who must be on it: Bill Layman.

Layman was a veteran engineer, one of the most senior and respected at the Laboratory. He had personally designed hardware for JPL space-craft that, except for Pluto, had explored every planet in the solar system. Layman had, over the years, gained a wide reputation for being an un-prejudiced mediator. He once recalled being asked to manage a task that had been going nowhere fast due to inter-organizational squabbling. "The turf wars were all at the high mucky-muck level, so I just gathered the troops around and basically said, 'Here's this really hard problem. Let's figure out how we can solve it.' And people pretty quickly forgot which division numbers they had stamped on their foreheads . . . In the process I got the name in some division managers' eyes, I think on both sides, as be-ing a neutral party. I don't really care a lot about division politics so long as I've got an interesting job. So I think, based on that experience, the di-visions decided that they needed someone to add to the brew that was be-ing managed by Donna Shirley."

Layman and Shirley talked about how they should split up the job be-tween them. "We both tried to outline where we thought we were strong, what we might be able to bring to the party," Layman recalled. "We agreed that there was plenty of work for the two of us, why didn't we just not write down a role statement for the next few months, and just take the work as it came. When one person was too busy, the other person would step in, and vice versa. You always worry about a contract like that turn-ing into a nightmare—but it actually worked out just the opposite."

The other engineers working on the rover effort knew nothing of management's imposition of Layman onto the team. Shirley introduced Layman to the core team as a great mechanical engineer, perhaps the best engineer of any type at JPL. Shirley made it clear that the team was lucky to have him as the microrover Chief Engineer. The team members who had worked with Layman in the past knew that was true; the rest of the team soon learned it themselves.

Over time the roles of Bill Layman and Donna Shirley became clear. Shirley had a hand in the technical design of the rover, but she reigned supreme in fighting the managerial and political battles that swirl around every project. Bill Layman was the engineering problem-solver and the source of technical leadership and vision for the rover team. Someone on the rover team put it this way: "Well, we go to Layman with the bad news, and Donna with the good news."

As the team of engineers who would build the flight microrover began to assemble, Dave Miller and his Robotic Intelligence group were nowhere to be found. In the fight over who would control rovers at JPL, Miller's section had lost. During the Mars Science Microrover effort, the Robotic Intelligence group had been transplanted into the same section as the Robotic Vehicles group. Miller and Desai were hostile to their new management, and by the end of MSM were looking for a way out.

When the charter battle was lost, the Robotic Intelligence group had been directed to move its offices to Building 107. Miller and Desai balked, pointing out that the facilities in 107 were not nearly as plush as their current offices. In response to Miller's complaints over the unsuitability of the building, JPL management broke free construction funding to replace the existing cubicles on the second-floor "penthouse" with walled offices.

But Miller and his crew would never occupy those new offices. By the time the paint was dry, Miller had gone on sabbatical from JPL, never to return. Rajiv Desai and the other members of the Robotic Intelligence group migrated back to their old section, promising to stay out of the rover business. Many of them would have been accepted into the flight rover team, if they had wanted to be part of it. But most of the Robotic Intelligence team consisted of researchers intent on proving out new algorithms. They lost interest after the basic principles had been demonstrated.

Yet Miller's group had made very real contributions to the Mars rover mission. The team had developed the software and control electronics for the Mars Science Microrover. Perhaps most important, Miller had personally promoted the idea of small robots, and redirected the focus of JPL, NASA, and the science community away from approaches that were too grandiose to be realizable implementations for Mars. Without the new microrover mind-set, there would certainly have been no rover on MESUR Pathfinder, and perhaps no Pathfinder mission at all.

And several engineers in the Robotic Vehicles group were very happy to accept Miller's legacy of walled offices on the second floor of 107.

In July of 1992, Shirley began holding weekly "core team" meetings, every Monday morning. Early on, the attendees included the usual suspects of JPL rover research: Don Bickler, Brian Wilcox, and other group supervisors with experience in the relevant technology areas. Then the Cognizant Engineers for each of the subsystems came on board, and the supervisors faded into the woodwork. The Cognizant Engineers had ultimate responsibility for the success of their subsystems. In the JPL environment, where most employees worked multiple tasks in parallel, often charging to four different accounts, the Cognizant Engineers would devote their full energies to only one job.

The Cognizant Engineers would staff their own subsystem teams. As much as possible, they would try to bring engineers onto their teams full-time. But their budgets were limited and there were many specialized jobs to be done. Nearly two hundred people would eventually participate in the design and development of the flight rover. Yet even when the rover effort was in full swing, there would never be more than the equivalent of thirty full-time employees on the payroll.

In the beginning, the rover design was up for grabs. Shirley's fledgling team studied all sorts of technology options for the new rover, just as her old Mars Rover Sample Return team had done. But we couldn't tarry, since we were on the hook to deliver a working Mars rover in three and a half years. "First it was like going through a catalogue and picking out your favorite goodies," said Bill Layman. "There were all kinds of things the technology tasks had developed, all the way from vision and mapping, robotic control, to mobility. Some of them you could adopt completely; others you couldn't hope to adopt given the constraints we had for size, power, and the schedule."

One of the big questions was "What will the rover's brain look like?" Rocky 4's brain for Mars Science Microrover would not do. A real Mars rover's CPU must be "flight-qualified" for the rigors of space. There were people who wanted the rover's computer to be really smart, really powerful. The lead proponent of the "egghead" rover brain was an electronics group supervisor named Leon Alkalaj. Alkalaj looked at the flight rover as an opportunity to develop new computing technologies: advanced processors, multitasking operating systems. Flight projects were where the big money was. Pathfinder's rover should have the resources to implement the high-performance computing research Alkalaj had been pursuing. Soon he learned from Bill Layman the real constraints of this new low-cost flight project. As Layman recalled, "I had by that time become well enough acquainted with how much power was available, and in previous spacecraft designs I had run into the problems associated with speed . . . Leon was the key player at that point. I basically put in front of him that the thing should draw no more than three-quarters of a watt, and I didn't care how dumb it was, that we would just slow down the rover as much as was necessary to make it think slowly. As long as it would think clearly, it would be okay." Three-quarters of a watt? That was nothing! Layman was telling him, "Either burn my strawman or adopt it." It didn't burn very well. Alkalaj had no alternatives to offer that could both survive the space environment and draw so little power.

Layman's constraint derived from the size of the rover. The flight rover would be no bigger than Rocky 4, and would be solar-powered. The small solar array that could fit on top of the rover would generate only a few watts of power. Those watts would have to be enough not only for the rover's brain, but also for its radio, sensors, instruments, and motors.

To find a CPU to meet this severe power constraint, the team reached into the past. The 80C85 microprocessor was low-power and rock-solid in the face of the radiation environment of space and Mars. The 80C85 design was already about twenty years old. It was *slow*, with perhaps a tenth the speed of a typical desktop computer then available. But it would do what the rover needed to do.

What if a better CPU showed up sometime later? To keep the options open, the electronics that interfaced to the onboard sensors and motors would be built on a separate circuit board; the CPU board could then be swapped out if a new alternative became available.

Unlike most JPL spacecraft, built with duplicate components for reliability, the rover would be "single string," with only one rover brain, one radio, one set of sensors. The usual deep space project would employ engineers to analyze every possible way the mission could fail as a consequence of the failure of a single spacecraft component. The project engineering team would then design out these "single-point failures," ensuring that only the simultaneous failure of at least two related parts could kill the spacecraft. This approach made JPL spacecraft very reliable. But it also made them bigger, heavier, and more expensive. Pathfinder and its rover were supposed to be "faster, better, cheaper." We couldn't afford a fully redundant rover: It wouldn't fit and it would cost too much. So the new single-string rover would survive only as long as its weakest component.

During the summer of '92 the rover team struggled with all of the major design questions. Should the rover be six- or eight-wheeled? Should the onboard batteries be rechargeable by the solar array or not? How would we keep the electronics and batteries from freezing in the cold Martian night? What sort of radios would be best to communicate between the lander and the rover? What kind of cameras should be mounted onboard? Would there be cameras at all? Should the rover use CARD or behavior control? How should the vehicle detect and avoid hazards? How much improvement in navigation performance over the research rovers would we need to satisfy our mission objectives?

What was the microrover's mission to be? Just as MESUR Pathfinder's purpose was to prove that the MESUR mission concept of small cheap landers was feasible, the rover would be a "technology experiment" that would show the utility of robotic rovers for future Mars missions. Most deep space missions were science-driven: Their design was molded to satisfy the requirements of the relevant community of scientists and the NASA-selected Principal Investigators. MESUR Pathfinder, and the rover that would ride on it, was a "technology mission": Science requirements would not be permitted to drive up cost. On the contrary, Pathfinder would have a mandate to cut capability if necessary to keep costs within the specified budget.

The rover's mission started out looking like a near duplicate of the Mars Science Microrover demonstration. Once off the lander, the flight rover would deploy a micro-seismometer, navigate across the terrain to a rock, chip away at the rock (we hoped more successfully than Rocky 4!), and then, using an Alpha Proton X-ray Spectrometer instead of Rocky 4's visible point spectrometer, determine the composition of the rock. Along the way, the rover would take pictures. Some of those pictures would document the condition of the MESUR Pathfinder lander, which might show some dents after a rough landing. And just by commanding the lander to image the rover during the rover's travels, we would learn about the soil properties of the Mars surface, and how well the rover performed.

Donna Shirley shrewdly declared that the rover mission would be only seven Martian days in length. Promising such a short rover lifetime had two major payoffs. The overall design of the rover could be simplified, which would help keep the cost down. And since the rover had a good chance of surviving beyond seven days, the probability of mission success was high. The rover team could declare success early and then keep going. Each additional day of rover operations would be another day exceeding expectations, pure gravy.

* * *

The first sanity check took place in October 1992. It was called the "Red Team Review." In NASA parlance a "red team" was a group of technically astute but disinterested engineers empowered to cast a cold eye on the early stages of a new project. Their job was to assess whether the rover team could get from here to there for the money available.

The Cognizant Engineers each presented their pieces of the rover design.

Bill Layman described the eight-wheeled "rocker-double-bogie" vehicle design he was partial to. The eight-wheeler was an even more capable variant of Bickler's rocker-bogie, and would work equally well driving forward or backward over obstacles. While Layman currently preferred this design, he made it clear that the team was still considering tradeoffs between four-, six-, and eight-wheeled concepts.

The rover power source would consist of a solar array and non-rechargeable batteries. Rechargeable batteries would have been eighteen times heavier for the same storage capacity! And rechargeable batteries would not operate below freezing, making them much less resilient to temperature fluctuations than the other electronics in the rover. For a short mission, nonrechargeables would support all necessary nighttime operations and emergency power needs while driving, while continuing to function at the lowest temperatures.

Layman figured the reviewers' going-in assumption was probably "The rover couldn't possibly work!" He wanted to convince them of the reliability of the rover's design. So in his presentation, Layman pointed out a key feature of the power subsystem. He had insisted on putting enough batteries into the rover so that even if the solar array failed completely after landing, the rover could survive and operate for a full week before exhausting its energy. And if instead, by some catastrophe, the rover's batteries were drained on the way to Mars, the rover could still complete its mission objectives using solar power only. The rover might be a low-cost system, but his team was thinking like a spacecraft design team.

The red team listened to the presentations for a full day. It asked questions and wrote recommendations.

The reviewers did not believe the design the rover team showed them could be done for the money. But they did believe that there was a rover that was doable for the available budget and schedule, if the rover team limited the requirements they were trying to satisfy. "The requirements must be brutally prioritized." The red team instructed Shirley's group: "Go with what you know how to do *right now*." The Rocky 4 design already existed: Use it. Forget the eight-wheeler. Forget future options for the rover CPU: Commit to the 80C85. The rover effort was fraught with uncertainty: Maintain large monetary reserves—perhaps as much as 50 percent—to manage the resultant risk. The red team didn't like Donna Shirley's seven-day mission. The reviewers preferred thirty days. They didn't want the rover to do more; they just wanted to give the future operations team more time to complete the same mission objectives. That also meant the rover could be dumber in its navigation across the terrain, almost as dumb as Rocky 4.

The message was: Simplify the design. To the red team, we were still trying to accomplish too much with too little.

The rover team listened. Mostly.

✳

Other reviews followed. We all hated these interruptions to the "real work" of engineering the rover. But the reality was that this series of reviews early in the project often put a spotlight on issues that begged for solution, but which the rover team did not have the authority to resolve on its own.

In one case, the Pathfinder science integration team had grandiose plans for the rover, loading it down with over nine pounds of science instruments plus the weight of the devices that would deploy those instruments from the rover. Their proposed science payload for the rover included not only the APXS and a micro-seismometer, but a neutron spectrometer intended to search for water. For a rover whose total mass could never be greater than about twenty-five pounds, these instruments would be quite a ball and chain. Worse, the details of the instruments hadn't yet been defined. The science guys were implicitly telling the rover team to stall its design effort until they made up their minds. At one of the reviews, the rover team presented the instrument mass requirement they thought the rover could handle: 2.6 pounds. The science representatives objected and presented their own number. The review board declared the science requirement unacceptable. Within a week, the proposed micro-seismometer was gone. The neutron spectrometer was gone. All that remained was the APXS. The design could move forward.

✳

While the rover had been funded for some time, the mission it was intended to fly on had not yet received final approval from Congress as part of the NASA budget. Once that was taken care of, Pathfinder would be nearly assured of full funding to completion.

And then Mars Observer, one of the most visible JPL missions, launched months before, suddenly disappeared. It was loaded with instruments, including a camera that would take images of the Martian surface with better than six-foot resolution. The $500 million spacecraft had been just about ready to be nudged into orbit around Mars. JPL operators had sent the commands to prepare the spacecraft. Part of the command sequence involved shutting down the communications transmitter, then turning it on again. But after the sequence began, there was only silence. For days, attempts were made to send new commands to the spacecraft. The antennas of NASA's Deep Space Network (DSN) listened in vain for any signals Mars Observer might be sending.

It seemed that NASA's string of failures that began with the Challenger disaster had not ended. This one pointed straight at JPL. Would Congress view this as a sign of incompetence, a lack of worthiness, and vote down JPL's newest deep space mission?

For days I was agitated as I contemplated my uncertain future, with Mars Observer in the news and MESUR Pathfinder on the chopping block. Would I soon be looking for work? In the end, despite the Mars Observer mystery, Congress approved MESUR Pathfinder. Those of us working the mission and its rover payload collectively relaxed. Now all we had to do was get to Mars.

EIGHT

THE ROVER WAR

I n late August 1992, Donna Shirley and Bill Layman briefed Pathfinder Project Manager Tony Spear and the MESUR Pathfinder project staff on the current status of the microrover design. At this point, only a couple of months after the effort had begun, they had only a "conceptual design" to present. There were still many tradeoffs to be made among design options. One of those trades was that between a tethered or untethered rover. A tether would provide a physical connection between the lander and the rover: Across that connection could flow power, communications, and control signals for the individual devices onboard the rover. An untethered rover would have to provide for itself, carrying its own power source, communicating with the lander—and through the lander with Earth-based operators—via a radio link, and relying on its own computer for navigating over the alien terrain.

Spear favored the tethered approach; he said he feared that a radio link between the lander and rover might be unreliable. He compared the radio link on a moving rover to carrying a portable phone from room to room at home, and wondered if the connection would be as easily broken. As one engineer on Pathfinder put it: "Tony mistrusted the radios. Period. I mean, really mistrusted radios. He had had some obviously very bad experience in the past—I don't know what that was—and he saw the radio link as nothing but a rat hole for money. He also saw a situation where the rover would be puttering along and you'd have a line-of-sight problem with a rock between the rover and the lander."

Some of the engineers in the meeting thought that a tethered rover would be a simpler rover, and not just because the radio could be replaced with a direct "telephone line" to the lander. One of the big problems for such a small rover was keeping warm in the cold Martian environment. (It was much easier to keep a large object warm at night than a small one. For its size, a small object will have more surface area exposed to the cold than a larger version of itself.) Electricity from the lander could flow into the rover across the tether, supplying power to heaters that could maintain the rover's temperature overnight. And with the lander acting as power source, the rover would no longer need its own batteries. Perhaps the rover could be simplified even further: eliminate the rover's computer, and let the lander's computer send signals directly to each motor on the rover.

But there were problems with the tethered approach. The rover couldn't drag a hundred yards' worth of cable around with it as it moved: The tether would get tangled, would catch on rocks, abrade, and finally break. Instead, you would need a tether spool on the rover to play out the line as the rover drove along. If the spool got jammed, or the rover got tangled in the ever-lengthening line, then the rover's mission was over. The bottom line was that a tether would severely restrict the rover's ability to perform its primary function: exploring the surface.

Layman and Shirley recommended the untethered rover concept. Most of those in the room agreed. But Tony Spear was adamant about the tether. The friction between Shirley and Spear was obvious. It clearly went beyond the issue at hand. The two of them did not get along. By the time the briefing ended, it seemed that Spear didn't want any kind of rover on his spacecraft.

When Shirley brought the tethered option to the rover team, Henry Stone groaned. He had just come off of the Hazbot project, in which he had been adapting a commercial mobile robot platform to operate in hazardous environments. That robot had had a tether. Stone had personal experience with how difficult it was to "manage" a tether. It had left a bad taste in his mouth. He wanted no part of it.

The concept of removing the rover's computer from the design also did not go over well. Not only would it make the rover a mere appendage of the lander, it was not practicable. Sending all the signals needed to operate the rover's actuators along the tether would not be easy. Unless the tether was a cable as thick as your thumb, the signals would have to share the same wires, then be separated out on the rover side, and sent to the appropriate devices. The "separating out" would require sufficient electronics to be almost as complex as the currently planned rover computer would be.

All I could think of was how we would operate the rover with a tether trailing behind it. If it was a rule that the rover could never drive over its own cable (or risk getting the tether wrapped around an axle), then how could we hope to drive up to a rock, take measurements, and then back away from it? One of the important characteristics of the rover was that it could turn in place, so it could maneuver in tight spots. With a tether hanging out in back, we'd never be able to take advantage of that capability. A tether would be an operations nightmare.

The rover team knew in its gut that a tether was the wrong way to go. We thought Spear was crazy. We set out to prove that the untethered rover would work.

Layman viewed the tether fight as a wasted, lost year. How much more could we have done if the rover team had not had to expend its resources defending itself, but instead had devoted its time to improving the design? Much later, Henry Stone would have a more positive perspective: he considered this battle with Spear to be key in taking the people working on the rover and melding them together into a true team. The rover personnel now had a common enemy. We would work together or risk having the dream of getting to Mars perish in its infancy.

＊

A few months after the conclusion of the Mars Science Microrover demonstration, Tony Spear had hired Lonne Lane as the Pathfinder Science Office Manager/Rover Manager. Lane would be responsible for coordinating the development of the Pathfinder science payload: the lander camera, the APXS, the neutron spectrometer, and any other instruments that might be selected. And both Spear and Lane assumed that Lane would manage the flight rover development activity. But when Lane returned from a Pathfinder-related trip to Russia, he was amazed to discover that Donna Shirley had assumed the role of Rover Manager. "Tony had his view of what was to be done. And Tony also wanted to know why I never received indication of an ad for such a position. He really wanted to know why. The answer is: I never saw the ad. Donna told me later it was publicly out. But if it was, I wasn't in the country. And you just don't do those kinds of things. So, the answer is, if it was a competed position, I sure didn't know about it."

Pathfinder and the rover were funded out of separate pots of NASA funds. The Pathfinder project received its funds from the NASA Code S organization, while the rover was supported out of the research-oriented Code R. Due to this organizational detail, the rover was not truly part of the Pathfinder project, and Spear did not have any authority over the rover funds. He had hoped that the joint appointment of Lane would put the rover and instruments together under Pathfinder management, ensuring coordinated development. Now it was clear that Shirley intended to manage the rover far more independently of Pathfinder than Spear had imagined.

Spear had a daunting job ahead of him: He was assigned the task of doing a Mars landing mission for one-fifteenth the cost of the last attempt: Adjusted for inflation, the Viking mission of the 1970s had cost about $3 billion. Spear had only $171 million. Although JPL had accepted the challenge of the "faster, better, cheaper" Pathfinder mission, the sense within the Pathfinder team was that JPL's upper-level management privately believed the mission could not be done. Spear was nearing the end of his career and was willing to take the risk. But he well knew that if Pathfinder were to have a hope of succeeding, it would have to be an extremely well-focused effort. Beyond the natural friction that existed between him and Shirley, Spear worried that efforts to incorporate the separately funded and managed flight rover would divert energies away from the design of the lander, which was the heart of the Pathfinder mission. He could ill afford such distractions. And who knew if Shirley's rover would do what the mission needed?

So in the middle of July, Spear went to Lane and his system engineer Bob Wilson. As Lane remembered the meeting, Spear said, "Give me a cost estimate for a derivative of the science rover that you did, the demo. I want to know what's the cost to do one or two of 'em." Wilson and Lane complied, putting together an estimate for two rovers that would deploy an APXS spectrometer to figure out what rocks were made of, would carry cameras, and might carry one more instrument as well.

Their first estimate came in too high, if they wanted to capture the Code R funds. There was a hard limit on the cost of the rover: $25 million.

Spear came back to Lane and asked, "Look, what's another approach? Be creative. What can you do if you want to get the science back?" Lane and Wilson mulled it over for four or five days. Then they said, "All right."

And then Spear said: "I want this in less than six weeks."

Lane nosed around NASA's Code R to find out if they would be open to an alternative to Shirley's rover. He made a few phone calls to people he had gotten to know during the Mars Science Microrover effort, which had been partially funded by Code R. When he got them on the phone, Lane asked, "Would you consider other approaches beyond or different from what is being pursued on the existing microrover?" Their answer was another question: "Like what?" Lane could only say, "I don't know yet. We'll call you back in a few days after we mature some ideas."

Lane and Wilson went to work on the rover study. "Given six weeks, we assembled a four-person team." They pulled in a couple of engineers from two aerospace companies they had worked with recently on other projects. By early September, they would have to be finished.

✳

Being creative often means eliminating as much complexity as possible.

Lane's team started with a small rocker-bogie concept, then simplified it to a four-wheeled design. "It was projected that the area around the lander was going to be quite benign, really," Bob Wilson said. "So, trades to save money by minimizing the capability of the rover were definitely made. And that's where the rocker-bogie assembly went away. It went to four wheels. We weren't going to do rock climbing, period. That saved a lot of money."

Then Lane and his team looked at the possibility of a physical connection between lander and rover. "The tether idea came along. We could do things that were very clever. If you did a tethered system, you could deliver power. That made the thing a lot smaller. You could deliver data back at a very high rate, if you so chose." You could use the lander computer to do image processing. If it got cold at night, you just sent power down the line to operate heaters inside the rover.

Lane's team sat down with engineers at Hughes Aircraft, which had developed tethers for wire-guided missiles. The tether they proposed for the four-wheeled rover was a direct derivative of this technology. There were ways of wrapping a tether around a spool that would reliably release the line without jamming. As Wilson commented, "When you were out of tether, you were out of mission." But who cared? The tether was so thin that it could be miles long.

Unlike Shirley's team, Lane's group focused solely on satisfying the anticipated science needs of Pathfinder. Lane said, "I will create to satisfy a need. We saw a science need. We thought we were being quite clever about bringing as much muscle as possible—scientific muscle—to the table for a finite amount of dollars. And we thought that was the charter we were given—to do a good science job on this mission. We wanted science return. You land, in a limited radius, what can you do that's really meaningful?"

The four-wheeled rover wouldn't have a computer onboard, except for those built into the individual instruments it would carry. The only rover computing would be on the lander, using the same computer the Pathfinder lander relied on to operate. Lane was proud of the new concept: "It was not like a Rocky, but it was able to get the instruments out, and do something with them."

＊

Soon after the study began, Lane realized he had a big problem: "Into the third or fourth week, it was very clear that Code R had no interest in this." Code R had been investing in rover research for years, and was now, for the first time, funding a flight rover. They wanted the rover to be the culmination of all of that research, an exciting practical application of the new technologies that had arisen from Code R funding. They made no secret of their intent for the flight rover to be a technology demonstration, not a science payload; engineers were already busy defining technology experiments to validate the rover's engineering performance once it was driving around on Mars. Bob Wilson described it this way: "There was a desire for an aggressive rover doing aggressive things. Our approach was to support the science: The science doesn't need to do aggressive things." This was the fundamental disconnect between the Code R research agenda and the objectives handed to Lane's study team by Tony Spear.

So midway through the rover study, Lane went to Spear and told him, "We think we have an interesting concept that probably would be viable. It would do the science. It would be an active rover out on the surface, that you could photograph and keep track of. It would have some unique capabilities. Yes, this is something we can do. I truly believe that, based on what experience we had with MSM. But I don't see Code R stepping up one iota to the line to support this. So the question is, if you want this, Tony, is Code S going to step up on the science side to support this?" But the money just was not there. So Spear told Lane to finish the study, document it, and close it out.

※

After the six-week study was finished, Wilson and Lane examined some smaller and simpler versions of the rover, but even this activity had trailed off to nothing by mid-October. To Lane, it seemed that the tethered-rover concept was dead. "It was quiet. Then all of a sudden we were notified that there would be a review of what we had done. That was about two weeks before it hit." So, just a few weeks after Shirley's rover team had presented to their own red team, Lane's tethered rover would face a red team review board as well. The red team for Lane's rover sought to compare his concept to Shirley's "free ranging" design.

Some of the engineers from Shirley's team were in the audience, either to present aspects of their own design or simply to watch. Here was Lonne Lane, who had managed the Mars Science Microrover, which had sold the idea of a flight microrover, presenting a competing rover design. For those of us on Shirley's team, Lane was now the enemy. I sat there, ready to pounce on any weakness in Lane's presentation.

Lane didn't even call his concept a rover, but the IDM, or Instrument Deployment Mechanism. He described its features: the tether, the instruments it would carry, the simple navigation concept.

If the review board had been skeptical about getting Shirley's rover out the door for its $25 million budget, they were more leery of the proposal to build the Instrument Deployment Mechanism for the promised $17 million. They asked Lane how much contingency he had included in his budget and schedule.

Lane told the review board that he was assuming that his extremely able team of engineers would be working sixty or more hours a week to achieve his proposed schedule on time.

Everyone knew that the engineers on flight projects typically worked long hours to meet schedule. But those extra hours were something a manager held in reserve for unexpected problems. If you started a project already dependent on those long hours just to handle the problems you could anticipate today, you'd burn out your team before the surprises even arrived. They'd have nothing left to give when the crunch came. The review board wanted to see more reserves in Lane's plan.

When Lane prepared to bring in one of his non-JPL team members to discuss detailed technical results of the IDM study, the review board chairman vetoed the presentation on the grounds that the review contained JPL discreet information, and no outsiders would be permitted in the room. In spite of Lane's protestations, the chairman held firm. Lane attempted to limp along without one of his key technical contributors present.

Representing Shirley's rover, Don Bickler presented the impressive performance of the rocker-bogie, including its ability to drive over rocks higher than its own wheels. The review board wanted to know how the IDM compared. Bob Wilson felt unprepared to respond: Prior to the meeting, "we were led to believe that this was going to be a decision point based on the overall merits of the effort—the effort being the rover and all the data it would collect. So we put together our presentation addressing that." Lane and Wilson were shocked by the review's focus on mobility capability. "That's all it was decided upon. We were ambushed." Wilson shook his head. "It was six weeks of very, very hard work, incredibly long hours, totally wasted."

The engineers on Shirley's team who sat in the audience at the review knew nothing of the specific technical objectives that had guided the design of Lane's Instrument Deployment Mechanism. We judged the IDM by the same technical requirements we were designing our own rover to meet. We felt a Mars rover should be designed for rough terrain, and that cutting corners to simplify the system was a mistake. We didn't know that there was no money in the Pathfinder budget to fund the IDM. We thought the project was trying to fund its own rover, just so it could have total control over its development, and to cut us out. We thought that the future of our rover depended on this review. We didn't realize that the fight over Pathfinder's rover had largely been decided in our favor due to events that had already occurred far beyond the walls of this conference room: Code R was funding our rover because it would deliver the technology capabilities they had always supported; they would never pay for an IDM that served only the needs of science.

✳

The IDM might be dead, but Spear still feared a failure of the rover radio link. He continued to pressure Donna Shirley to put a tether onto her rover.

In January 1993, Carnegie-Melon University mounted a robotic expedition to Antarctica. Under NASA funding, Red Whittaker and his team had built Dante, an eight-legged walking and rappelling robot. Dante was intended to descend into its own "hell," actually the crater of the Mt. Erebus active volcano. With great effort, the CMU group had transported themselves and Dante from Pittsburgh to Antarctica, to the slopes of Mt. Erebus, and then to the lip of the volcano. They had hooked up the tether upon which Dante depended for power and communication. But after the robot took only a few steps, its tether snapped. The expedition was over.

At the next rover team meeting, Donna Shirley conveyed Spear's reaction to the news of Dante's broken tether: "Shit! We'd better go wireless!"

Two months later, Pathfinder engineers and scientists made a field trip to study the geology of Death Valley and the Mojave Desert for insights that might impact landing and roving. During the field trip, the rover telecommunications team demonstrated the proposed rover radios to a range of three hundred yards, several times farther than anyone imagined driving the rover out from the lander on Mars. At the conclusion of the radio test, Tony Spear gushed his relief, congratulating the rover communications Cognizant Engineer for demonstrating that the radios would indeed work better than his cordless phone.

NINE

A DESIGN THAT REALLY WORKS

By 1994 the pace of the rover design effort was accelerating. The flight rover effort had been christened with its official name: MFEX, short for Microrover Flight Experiment. The period of heavy reviews and threats to the existence of MFEX had passed. Donna Shirley was protecting the team from undue reporting requirements by declaring to the line management that she would be the sole source of information about team progress: Leave the rover team alone and let them do their jobs!

It built up slowly. No one could say exactly when the feeling became universal. But there came a time when the MFEX team felt like a train barreling down the track. From the many months of frustration, distractions, and apparent lack of progress, the team had emerged leaner, determined, and with the framework of the design in place.

There were lots of engineering problems to solve, but the team was handling most of them as fast as they were coming up. A rover engineer could look around and see the other engineers on the team pushing ahead. The momentum of the team drove the individual members into a run just to keep pace.

At the weekly core team meetings, the Cognizant Engineer from each subsystem would report the significant activities in his or her area for the prior week. Often a key issue would come up, one that affected more than one subsystem; the meeting could instantly segue from status reporting to a detailed design session. Many sticky problems were solved then and there, allowing the design effort to proceed. The meetings often ran over three hours, until the exhausted engineers dragged themselves off to lunch, only to continue the discussion as they ate. During the meetings Shirley constantly typed notes into her laptop computer, recording everything. Within a day she would send out an email of "Rover Significant Events" providing minutes of the meeting in excruciating detail.

The Pathfinder project and its science payload people often attempted to influence the design and features of the rover—the characteristics of the rover's cameras, which instruments the rover would carry, what experiments the rover's software would support. Sometimes these requests would travel not to Shirley, but directly to the engineers on the rover team. When informed at the weekly meetings of such requests, Shirley would remind her team of what she called Donna's rule: "A requirement is not a requirement until someone pays for it." Pathfinder had far more desires than money to pay for them. And if they weren't willing to pay, Shirley did not feel obliged to be responsive. Donna's rule was also a warning to the rover team to avoid expending effort to satisfy technical requests that, taken one at a time, seemed eminently reasonable, but together would ruin the effectiveness of the team.

The evolution of the MFEX system was following a pattern of major design problems uncovered and new challenges met. Each subsystem would deal with issues that could potentially doom the rover. The solutions to these difficulties would often pose new challenges to another subsystem.

Bill Layman kept the team moving. "I felt it was important to manage those technical problems so that people didn't just dither—go into hysteria—which is a natural reaction of a committed engineer if you just keep tightening the schedule screws on him, and not relieving him of any technical requirements. You finally reach a point where they really can't do it.

And *they* know that. And they don't know how to proceed. They stop being logical. The best engineers have the quality of saying, 'Well, these people are just nuts, setting these schedule constraints. I'll just start here at the beginning of this problem and work in an orderly way from the beginning to the end.' But if you tighten the screws tight enough, almost everyone will eventually become inefficient, because they try to figure some way to leap to the end. They stop being methodical about getting the job done." Layman watched for signs of overload within the team.

But there was also an upside to the stress. "There's a certain amount of that that's a payoff: If you force people to take bigger intuitive leaps, and their intuition is sound, then you can get more and more efficient as you force larger and larger leaps. There's some optimum point where they succeed with ninety percent of their leaps. For the ten percent they don't make, there's time to remake that leap or go a more meticulous route to the solution of that particular problem." The optimum level of stress? Layman laughed. "That's right. There's the optimum level of stress, where you're taking risks intentionally, to accumulate a budget of time and money, which you then spend to solve the problems where you missed your guess or the risk was too large. I felt like our team was operating at just about that optimal level of risk. The rover team was competent in every area . . . and overworked in every area."

Layman had to deal with the dynamics between rover team members as well. One day the Pathfinder flight system manager dropped by his office and asked Layman how his new job as rover Chief Engineer was going. Layman's answer: "The hardest part of the job is keeping Howard Eisen and Henry Stone from tearing each other apart! The rest of it is a piece of cake." Eisen and Stone were the Cognizant Engineers for the two biggest subsystems, and they were constantly battling over technical approaches and who had agreed to what. (Surprisingly to many others, the two of them partnered for sailing races after hours. Sailing together relieved stress and reminded them they were teammates.) The assembled team was full of strong personalities. They had different skills, different priorities, and distinct ways of doing things. But in the two most important characteristics, they were the same: They were all good at what they did, and they all wanted to put a rover on Mars. Those who did not share these two traits did not last long on the project.

✳

Layman's imprint was on every aspect of the rover design. It was there in how he got the most out of every member of the rover team. If an engineer came to Layman with a solution to a design issue that would work "most of the time," the Chief Engineer would often ask about possible ways to "kill the problem," overwhelming it so that there was no credible failure possibility left. Layman was insightful enough to know which problems had to be killed, and which could afford the risk. And when an engineer proposed an easy way out, Layman might ask him, "If, years from now, you stood up before a review board to explain why the rover failed, what would you tell them?" Layman called it "designing on the path of least regret."

Layman continued to be concerned over power on the rover: "Every spacecraft I've ever worked on had at least two crises where there was not enough power." You dealt with this sort of problem by either getting more power to begin with, or using what you got more efficiently. Usually both. The rover's primary source of electrical power would be its solar array, a flat panel that would cover the top of the vehicle. The onboard batteries were there mostly for emergencies. Layman wanted the solar array to get bigger.

Solar panels are fragile things. The mechanical team wanted to keep the panel out of the way, inside the footprint of the rover's wheels, so that a passing rock wouldn't shatter any cells. With a smaller panel, they also wouldn't have to worry about collisions between the panel and the rover's own wheels, going up and down on the rocker-bogies. The rover Chief Engineer pushed on the mobility team to make the panel as large as possible, extending the array over the wheels: "The solar panel needed to be as big as we dared make it." In the final compromise, the mechanical team cut out the front corners of the panel above the front wheels, just enough to keep the rover from injuring itself as it drove over rocks. And the control and navigation guys had to make sure their navigation system was good enough to keep the overhanging solar array away from the natural hazards of Mars.

Because the rover would have to get by on the limited power available from its solar panel, it was imperative that the rover brain supply power only to the devices it needed at a particular time. This necessitated including a series of CPU-activated power switches in the rover electronics boards. When Layman saw the inefficiencies in the power switching system first proposed by the electronics guys, he counterproposed: "Let's do everything with mechanical relays." He wasn't even sure his suggestion was possible, but he wanted to galvanize the team into coming up with a better design. Perhaps partly because the electronics engineers found Layman's mechanical solution so repulsive, they went away and then came back with a far more efficient solid-state design.

※

Layman defined the state of rover thermal control as it existed at the start of MFEX: "Rocky 4 had a sheet metal frame and a pile of circuit boards on top of it. Nobody had a vision about how that could be configured to survive the Mars environment. It was clear immediately to everyone that we needed what amounted to a beer cooler that we put all the sensitive stuff inside of and kept warm." The thermos bottle concept that would keep the rover electronics from freezing was quickly named the Warm Electronics Box, WEB for short. The WEB would warm up during the day while the electronics were on and generating lots of heat. Then, overnight, the thermos bottle would slowly cool off. By the time it dropped close to the electronics' lower limit of $-40°F$, it would be morning, the rover would wake up, and heating would begin again.

Layman presented yet another ugly solution to motivate the design of the WEB. He suggested a big box to hold the electronics, with lots of insulation to keep it warm. The box was so big that it would hang down low between the rover's wheels and was likely to bump into lots of rocks.

Howard Eisen recognized the design of the WEB as one of his mechanical team's main technical challenges. They set out to improve Layman's WEB concept into something practical, something that would work without degrading the rocker-bogie's mobility performance. They needed to make the box smaller. To do that, everything that required the thermal protection of the WEB—electronics boards, navigation sensors, radio, and batteries—would have to be packed closer together. Layman took the position that the mechanical guys would never get all the electronics into the tiny box they were imagining. "Show me," he said. And eventually they did.

The thermal analysis of the WEB design revealed yet another problem: The inside of the WEB was still going to get too cold overnight. No matter how much insulation you layered on, the inside of the WEB would eventually reach the average temperature of the Martian environment in which it sat. The rover needed another heat source, one that would stay on all the time. But they couldn't afford to exhaust the rover's battery to run heaters at night. The only option remaining was one commonly applied on deep space missions, but increasingly out of favor: Radioisotope Heater Units, or RHUs. An RHU consisted of a tiny plug of plutonium encased in a C cell–sized graphite container. A single RHU generated just about a watt of heat. The beauty of it was that it would continue pumping out that watt for years, with only the slightest degradation. There wasn't a battery on Earth that could come anywhere close to this capacity with the tiny mass and volume of an RHU. But RHUs were controversial. Although millions of dollars had gone into detailed studies over the years to establish the safe design and handling of RHUs, the public remained fearful of anything radioactive.

Eisen set about determining the feasibility and cost of putting three RHUs into the MFEX rover. He soon learned two facts. First, there were a number of RHUs already in existence, spares manufactured for the Galileo spacecraft, but never used. These RHUs were in the custodianship of the Department of Energy. Second, there was no mechanism for charging a new project for the use of those RHUs. So, once MFEX completed all of the appropriate procedures, the RHUs would be free!

Eisen's team also had to find a good way to insulate the WEB. The standard methods of insulating equipment in space depended on the presence of vacuum—they counted on nothing being there. Vacuum inside the walls of the WEB would work fine on the way to Mars. But after landing, the tenuous atmosphere would seep into the space between the walls, creating a "thermal short," like an electrical short circuit, that would allow heat to flow too freely out of the WEB. The only way to keep the Martian atmosphere out would be to put something else in. At first the WEB engineer settled on powdered aerogel, a lightweight material that was an extremely good insulator. You could pour the powder into the fiberglass honeycomb that formed the WEB walls. But when the batch of aerogel arrived from the manufacturer, it was a third heavier than promised. The rover team had been desperately trying to stay within the mass allocation. Now we were in trouble.

Within weeks, Dave Braun, an engineer on Eisen's team, had a conversation with Peter Tsou. Tsou was a JPL engineer with a facility that could manufacture small quantities of an alternative form of aerogel—a solid form. Braun thought that solid aerogel just might do the job of insulating the WEB. But there were problems with this notion: If you tried to cut small pieces of solid aerogel and insert them into the honeycomb, there would be air gaps. Those gaps would form more thermal shorts. A third engineer, Greg Hickey, suggested redesigning the WEB wall—don't use honeycomb material at all. He proposed a sheet-and-spar construction, similar to that of an airplane wing. There would be space for aerogel between the inner and outer WEB walls, except where fiberglass bulkheads would link the walls at regular intervals. The solid aerogel had to be made in slabs of precise thickness (the aerogel would crumble if you tried to machine it). To ensure that there were no air gaps in the walls, Hickey would build the walls around the slabs of aerogel, with the walls pressed tightly against the aerogel, compressing it.

Hickey assembled a sample WEB wall. It worked!

When I held a piece of aerogel in my palm, I was prompted to call it "solid air." The piece was a couple of inches on a side and yet so light that I couldn't feel the weight of it. I looked through the material to see my palm only slightly obscured by a smokiness in the aerogel. One of the engineers on the mechanical team had his own name for solid aerogel: "manna from heaven."

Using solid aerogel, the mass of the insulation inside the WEB walls dropped by over two pounds. For a rover that massed all of twenty-two pounds total, this weight savings was tremendous!

Sometimes the most mundane elements of the design would prove to be the very ones that required the greatest ingenuity to be made to work. One such area was the rover's wiring.

All of the connecting wiring among all of the boards, sensors, and motors on the rover was collectively referred to as the "cable harness." Early on in the MFEX design effort, there was no one responsible for the cable harness. But as the design began to mature, Layman knew that the time had come to find an engineer to oversee the rover's wiring. After a brief search, Layman brought Allen Sirota onboard the rover system team.

The last time I'd seen Sirota, he'd been leaving the Robby team to return to his first love—a flight project. At the time, a few years earlier, I'd shaken Sirota's hand and told him that if we ever built a Mars rover, we'd bring him back. Now I shook his hand again. Through coincidence and luck, we had made good on that promise.

Soon Sirota was the master of the rover interconnect diagram. This diagram didn't show the details of the computer boards; instead, it showed how those boards would connect to everything else on the rover. Wires led to every device. The APXS electronics, the rate sensor, the accelerometers—all would need power and data lines running to them. One cable started on the boards and split into several connectors that mated to connectors on the internal bulkhead; from here, cables ran back to the other devices inside the WEB. Another cable emanating from the electronics boards snaked through a tunnel to the outside of the rover to power and control the motors, cameras, and sensors that would be sitting out in the cold Martian breeze. As the design of the MFEX rover progressed, the wiring diagram went through many revisions, each more elaborate than the last.

You couldn't just string wires from one component to the next, soldering them at each end. Engineering must go into ensuring the rover could not just be put together, but taken apart as well. Once assembled, the rover would be subjected to tests, then inspected. Components might fail and require repair, or entire assemblies might be swapped out. Over time, subtle design flaws might be discovered, requiring modifications to the electronics boards. So removable connectors were good: They allowed the design to be more modular. But they also added mass and took up space.

The number of wires was starting to look like a problem. By the time the design stabilized, there would be 243 wires leading from the electronics boards out to the sensors and motors on the rover. So many wires would form a thick bundle, as thick as your thumb. This would be very difficult to handle; it would be too stiff to bend very much, so every time you placed the bundle down inside the rover and installed the boards, there was a good chance you'd break a wire. Then you'd have to repair it, and there was a chance you'd break other wires in the process.

And suppose you managed to install all that wiring without damage. Mars was *cold*. All those wires leading outside into the frigid Martian air would be even stiffer and more subject to breakage than on Earth. The rover was a complex set of moving parts. Many of the wires would need to flex as the rover traversed across the surface. How long would they last?

Worse, the copper that made up the wires was not only a conduit for power and data. It also conducted heat. The purpose of the rover's WEB was to keep the electronics and sensors inside warm despite the extreme cold of the Martian environment. But with so many wires routed from inside to outside, much of the WEB's warmth would be leaking out through them.

The mechanical team's solution to the heat leak was to create a cable tunnel, often called the "igloo tunnel" on the front of the WEB. The tunnel was a mini-labyrinth through which the wires would be routed. The convoluted path the wires would follow meant that the cables would be much longer, several feet, and the warmth of the WEB would flow much more slowly along the wires into the deep cold. The igloo tunnel was a good solution, but it also required the cables to make hairpin turns inside the labyrinth. How would the cable bundle reliably survive these tight bends?

Something in the design was going to have to change.

The issue stayed on the table for many months. Eventually, John Cardone, the designer doing the mechanical layout of the cabling, offered up a suggestion. He recommended using "flex-cable" technology, which would eliminate all of the separate wires, integrating them into a single ribbon cable. A ribbon would bend easily in some directions, almost not at all in others. So long as you anticipated the directions the cable had to flex in use, you could design a flex-cable that could be bent almost in half without stressing the "wires" inside it. Flex-cables were manufactured like circuit boards, with alternating layers of conductive "traces" and insulation. A flex-cable for the rover might require twenty-five or more such layers. Cardone had designed layouts of flex-cables for other projects. He'd never seen one as involved as what the rover would need, but there was no reason, in principle, that it couldn't be done.

Sirota and the rover team studied the idea. JPL did not have the facilities to build flex-cables in-house, so going with flex-cables would mean contracting them out. And a complex manufacturing process would cost money. But in this case the money seemed like a good tradeoff against the risk to the rover's development schedule if they went for a cheaper, unreliable design. After evaluation and debate within the rover team, the flex-cable approach was in.

Sirota had just hired Art Thompson onto the MFEX system team. Sirota and Thompson had worked together on Sirota's last assignment, a Space Shuttle flight experiment. Sirota immediately assigned Thompson to be the contract manager for the flex-cable procurement.

Pioneer Circuits got the contract. There would be three types of flex-cables, one inside the WEB between the two electronics boards, one from the boards to the internal bulkhead, and one that went through the igloo tunnel. The tiny rover's igloo tunnel flex-cable, as designed by Sirota and laid out by Cardone, would be the longest and most complex flex-cable Pioneer had ever produced. From one ribbon coming off the boards, the cable would split into six ribbon "fingers," each to be soldered to a separate connector. The sets of flex-cables would be expensive: Between the flight units, spares, and test runs, the contract was worth several hundred thousand dollars. But it should be worth it, creating a high-reliability link among the rover's electrical assemblies.

Not everything went according to plan.

When the first set of flight cables were delivered, Sirota began a careful inspection. The cables had been delivered without connectors; Sirota had separately purchased flight-qualified connectors, which would later be affixed to the flex-cables by experienced JPL flight electronics technicians. As part of his inspection, Sirota needed to confirm that the proper "wires" in each finger of the cable aligned with the appropriate solder location on the back of the connector. He placed the first connector against one of the fingers of the flex-cable. It didn't match. Sirota quickly checked the other connectors and fingers against each other. They didn't match either.

Sirota went pale. Something was fundamentally wrong. The arrangement of the wires in each finger of the flex-cable was backwards, a mirror image of what it should be! They had just spent $450,000 on the contract, with most of that pulled out of reserves. And the cable was useless!

It was instantly clear to Sirota how the error had occurred. Each finger of the flex-cable had been made to be soldered to a specific type of connector, called a "micro-D." They were called D connectors because, when you looked at them face-on, they had the shape of a tall, skinny letter D. D connectors had two genders, male and female, that fit together. Male connectors possessed pins that plugged into the female connectors' sockets. Micro-Ds had been chosen to minimize weight and volume: They were the smallest and lightest D-type connectors available. Their small size forced the pins to be recessed, so they looked a lot like sockets. And the sockets looked like pins. It was easy to get them mixed up. Sirota carefully checked over the JPL drawings. The error was there, a reverse interpretation of his intent. Sirota blamed himself: He hadn't personally checked the drawings that had already been reviewed by two other engineers. Pioneer Circuits had simply complied with the specifications in the JPL drawings and designed for male micro-Ds instead of female micro-Ds.

Other than this one huge error, the implementation of the flex-cables was perfect. What was he going to do?

The obvious choice would be to completely redo the design, which would basically mean starting over. ". . . and the timing. The timing [of the rover schedule] was so exact at that point that it would just be impossible." There was no time left to do the flex-cables over again. Those were the things bouncing around in Sirota's head. "What helped me eventually was realizing that the flex-harness would have to stay the same. Everything else would have to be reworked." How hard was that going to be? "I had the realization that most of the stuff hadn't been wired yet, and the stuff that had been wired could be reversed. Then I felt a little better. We could live through this.

"Basically, we reworked nine cables here, and told everybody who was still going to supply their cables to reverse them, because they hadn't wired them yet. We caught it really quickly. It could have been a lot worse. That was something maybe not many people knew about. It could have been a disaster, but that was averted very quickly by some fast thinking." Sirota was silent for a number of seconds. ". . . That's called dodging a bullet."

✵

Small though the rover was, it was still too big. Like all of the Rocky vehicles before it, the flight rover's rocker-bogie running gear gave it high ground clearance to safely traverse over rocks higher than its wheels. But this feature also made the rover taller. It was too tall to fit inside the Pathfinder lander during the trip to Mars. And all that empty space under the rover and between its wheels would have been wasted volume while it was sitting on the lander. Somehow, the mechanical team would have to find a way to make the rover crouch down while inside Pathfinder, then rise up to its full height once on Mars.

Howard Eisen had selected Ken Jewett to be the mechanical engineer responsible for the overall configuration of the rover. So figuring out how to make the rover stowable fell onto Jewett's shoulders. Where Eisen was loud and combative, Jewett was quiet and self-effacing. He just wanted to do the design work, solving the fundamental questions that stood in the way of making a mechanical system function: "The most creative part is in the design. The rest can get excruciatingly boring." Jewett didn't usually make a lot of noise, but he did have a temper. He didn't like it when other people tried to make their technical problems into his problem, due to laziness or lack of good engineering skill.

Stowing the rover was his problem. Jewett struggled with it. He could see that he would need to "break" the rocker. If you looked at the rover from the side, the rocker was the bigger of the two pieces of the rocker-bogie. The rocker had a wheel at the rear end, and came forward until it attached to the bogie at the front; the bogie, with a wheel on each end, pivoted freely around the forward end of the rocker. The rocker itself pivoted around the arbitrarily named "jeff tube," an axle that went all the way through the rover's WEB from side-to-side. The only way to reduce the height of the rover was to break the rocker into two pieces. Then the rover's body could drop down until the bottom of the WEB nearly touched the deck. When the time came, some mechanism would force the rover to stand up, and those two pieces of the rocker would return to their original "unbroken" shape. The rocker pieces would then have to latch permanently into place, with no chance of slipping back out. Otherwise, the rover's mission would be over; at its stowed height, the rover would be a low-rider with so little ground clearance that it would be defeated by the smallest rock.

How should the critical latch be designed? Jewett studied the problem, finding no easy answers. He left the latch problem alone for a while and worked his other rover design assignments: deploying the rover's antenna, mounting the cameras so that other components of the rover wouldn't get in the way, and generally just coming up with clever ways to make the rover lighter. But the latch was always waiting for him.

"That was a hard nut to crack. I kept putting it off. We tried several designs where there were linkages that came together, pins running in slots that would drop into a [notch] and couldn't get out of it." But with pins and slots there was always the chance something would bind, or a pin would get bent and stop working right. Jewett just didn't trust that there would be a sufficiently reliable version of these approaches that met the rover constraints.

"I fooled around with that design for about a year before I finally did an 'Aha!' and figured out how that would really work." That inspiration was to use a particular spring anchored in just the right spot to each of the two pieces of the folding rocker. "It's just a bent spring and as it comes up, it just *snaps* into a locked position. And it's very strong when it's finally in its locked position, yet it's very flexible before that." The final design was simpler than the earlier alternatives he had examined. "Simple is good."

Unlike the other designs, this one "just wanted to work." Once he had the design concept down, he could stand back and appreciate its elegance as something apart from himself, as if it were something discovered and not made: "Maaaaan, that's neat!"

※

Layman pushed on the mechanical team again, this time to find a secure way of mounting the rover on the lander deck. Not only did the body of the rover need to be held in place during launch and landing, but each of the six wheels had to be tied down separately as well. The team worked out a way to tie the rover down at three points, so there was no possible way the vehicle could accidentally come loose. Cable-cutter pyrotechnics, fired by the lander after it was safely down on the Martian surface, would release the rover. "Cowcatcher" hooks held each wheel in place by its cleats. Only the turning of the rover's wheels under power would release them from their restraints.

How was the rover going to get stood up so that the springs on either side of the rover would get the chance to *snap*? One of the early "standup" concepts was to put a big spring under the rover. As soon as the tie-downs holding the rover to the petal were released, the spring would push on the bottom of the WEB, lifting the rover upward. Layman did a calculation showing that any such spring would be so powerful that the rover would literally leap into the air, "as high as your shoulder," in the low Martian gravity. Not a good idea.

Eisen came back with a new idea: Let the rover stand up by itself. If the rear wheels drove forward, and the rest of the wheels stayed in their places, the rockers would start to lift. Keep driving those wheels, and eventually the rockers would lock. To the mechanical team, this solution was simplicity itself, elegant in that it involved no additional hardware or mass. But to the control and navigation team, it meant new software and careful testing. How would the rover tell that it was done standing up? If the wheels drove for too long, their cleats would break, and the rocker-bogies would be overstressed. There would be potentiometers on the bogies to measure how the bogies moved during traverses (How big a rock did we just drive over?); maybe the readings off the pots could be adapted to monitor standup.

Henry Stone wanted contact sensors on the rockers, small switches that would be triggered when the rockers locked. The switches would give a sure, positive indication that the vehicle had stood up. Potentiometers would have to be carefully calibrated, and might provide less certain results. Eisen complained that contact switches were not particularly reliable, added more mass to the rover, and required more wires going in and out of the WEB, which would contribute to the loss of precious heat from inside the rover. He proposed that the bogie pots would be good enough, and they would be there anyway. Eisen convinced Layman that the rover could do without contact sensors. Stone wasn't happy, but accepted the decision. The lack of those contact switches would eventually cause many headaches, and Stone would come to regret not having fought harder to keep them in the design.

✳

As the rover's design matured toward completion, events conspired to take our key leaders away from us. The Pathfinder lander design was in trouble. The project asked for Layman's help to get the lander out of the doldrums. The lander team was uncovering issues faster than it could deal with them. The lander was overweight, exceeding its mass allocation. There wasn't enough room for all of the subassemblies to fit in the available volume. Certain issues with the fundamental structural design of the lander were still not complete, and the clock was running. And it wasn't clear that the landing cushion airbags would survive if the lander came down with a significant horizontal velocity, which it could if the wind was blowing on landing day. Layman did not want to move off the microrover team, because he didn't feel he had finished the job. "I got pressured pretty seriously to leave the rover, and go try to fix the lander. I felt that the rover team relied on me, that I was an important part of the lubrication that kept the machinery turning in the team. My response to the management that was trying to move me wholly onto the lander was that they just couldn't do that. I couldn't abandon this team and expect it to function, because we had built a team that was one deep everywhere, and I was the 'one-deep' person in my particular spot, which was kind of the overview system architect." Layman was reluctant to shift his attention to the lander, but in the end felt he had little choice: If the lander failed, his rover would never get to Mars. He accepted the role of Pathfinder Project Mechanical Engineer as an additional duty piled on top of rover Chief Engineer. He knew that he would not be able to maintain a balance between rover and lander. The rover design activity was winding down, moving on to assembly and test. The lander was not as far along. He was sure to be sucked in by the lander, "inevitably drawn further and further away from the day-to-day doings of the rover."

Layman began passing pieces of his rover responsibilities to Sirota, Stone, Eisen, chief rover system engineer Jake Matijevic, and me. He hoped that the team would have a chance to transition gracefully. "I felt really guilty, like I had set up this organization, then backed out at a critical moment. But I could see no way to do anything else without essentially assuring that the lander would fail." As I learned why Layman was moving to the lander side of the house, I found it frightening that there were so many problems remaining on the lander at the same time that we were finalizing the rover design. Would the lander ever come together in time?

And then, in July 1995, Donna Shirley was promoted out of her job as Rover Manager. The number of Mars missions was growing. NASA wanted a whole series of missions, done in the "faster, better, cheaper" mold, with two spacecraft launches every two years. Responding to this new NASA mandate, JPL was creating a new internal organization, the Mars Exploration Directorate. Shirley was chosen as its director.

At an all-hands rover team meeting, Shirley announced that while she would try to keep a hand in as long as possible, Jake Matijevic would now be the official Rover Manager. She then said that the team would need someone with the same thoroughness and attention to detail as Jake to take on many of his duties. As we walked back from the meeting, I felt lots of eyes on me. Allen Sirota smiled and put a hand on my shoulder. "It's obviously you," he said. I wasn't so sure, but Henry Stone was looking grim, as if the decision had already been made. "I don't know what I'm going to do. Where am I going to find someone else to do what you do? I'm screwed." I was committed to Stone's control and navigation team. But I was also committed to the rover as a whole.

Within a few weeks, I'd been split in two, trying to do half of my old job for Stone and half of a new job for Matijevic.

TEN

THREE ROVERS

The control and navigation work on the flight rover was centered in Building 107: software development, sensor design, electronics and navigation algorithm testing. Henry Stone had moved his office into the building almost immediately after joining the MFEX effort.

My first encounter with Stone had been a few years earlier when he had heckled me at a talk I gave on Semi-Autonomous Navigation of the Robby rover. He had asked question after question about a detail of the algorithm. At first I had wondered if he was maliciously attempting to derail my presentation. I later realized that Stone was simply both incredibly detail-oriented and unusually persistent. He had an inherent need to delve into every aspect of a technical concept, and didn't feel satisfied until he understood every one of those details. This trait contributed greatly to his success as an engineer, although he sometimes forgot that others in the room were not necessarily vitally interested in such a complete analysis.

Stone seemed to be happy only when things were moving along at a fast pace. The other engineers he worked with, who respected him a great deal, sometimes kidded him about his quirky habits. One of those habits was absentmindedly pressing his hands against his temples so hard it seemed he was trying to crush his own head. It seemed to help him concentrate.

❋

Henry Stone smiled wistfully as he considered the first few months of MFEX: "I naively thought that we'd have such a small team that I would actually be involved in the software design." As the scope of the subsystem job became clear, and his responsibilities as Cognizant Engineer grew, Stone was forced to let go of this desire. Instead, he was pulled more and more toward the most complex hardware that would come out of the control and navigation subsystem: the electronics boards comprising the rover's custom brain. These boards would control everything the rover did, operating every device onboard.

Gary Bolotin joined the team as the lead rover electronics engineer. His job was to design the rover's computer. The meat of his work would not be the creation of new components, but the selection and arrangement of particular combinations of existing components into circuits that did specific useful things. There were circuits to switch the rover's motors on and off, and reverse their direction; electronics to read out sensor values to determine how far the rover had driven; an "alarm clock" to wake up the rover at the proper time after it had shut down to conserve its batteries; and on and on. The result of Bolotin's work would be schematics, virtually a paper representation of the thousand electronic components— and all of the interconnections among them—that would together constitute the brain of the flight rover.

The custom hardware that would embody that design and make it reality were the printed wiring boards. Within those boards would be etched all of the necessary circuitry. Once the boards were manufactured, every component would then be mounted and soldered into its reserved place. The volume constraints of the rover's WEB forced a hard limit on the size of the boards. The components would end up so tightly packed onto the boards that it would be a wonder if they all fit.

The flight rover couldn't use just any electronic parts that might be available commercially. Not all parts were built to the same level of reliability. There were standard commercial components, and "military specification" parts designed to work over wider temperature extremes. Most reliable of all were Class S flight-qualified parts, intended to survive harsh space environments. If a particular type of part failed during a space mission, an advisory would be issued warning of the risks associated with that part. There were engineers at JPL whose job was to make sure no such suspect parts found their way into new spacecraft.

Bolotin put together a spreadsheet to track all of the components he had incorporated into his design, parts that needed to be purchased. "It was the most detailed spreadsheet that I ever did—for anything." He gave a copy to Stone. The next time Bolotin saw the spreadsheet he had thought was complete, Stone had taken it and enlarged it enormously. He was tracking the status of every part destined for the electronics boards. For some unusual parts, small enough to fit on the crowded electronics boards, the estimated delivery dates were twenty-two weeks after receipt of the order. That was a five-month lead time!

*

How do you develop software for a flight rover that doesn't yet exist? "Rapid prototyping" is the process of building a practice version of something when you have barely the information you need to get started. Revealing problems in this way was the first big step to solving them.

Rocky 4's wheelbase was almost the same size as the flight rover's would be: As a rapid prototype, it would do. Howard Eisen's mobility team gutted Rocky 4, removing the electronics, instruments, and rock chipper used for the Mars Science Microrover demonstration. They replaced the existing wheels with wider-track stainless steel wheels, added prototype steering mechanisms and sensors, and then they drove this prototype rover through soils with a consistency that matched our best guess of Martian soil—fine and powdery like talcum powder. Once the mechanical engineers had satisfied themselves that the rover would perform just fine in alien soil, they turned their prototype vehicle back over to the control and navigation team.

Rocky 4 now became a dedicated testbed for exercising the software. Stone emphasized the point by calling the vehicle the SDM—Software Development Model. His team installed a small cardcage containing a commercial 80C85 CPU and wirewrapped electronics boards that together duplicated perhaps half of the functions of the future MFEX brain.

Accelerometers and a rate sensor measured the tilt and turns of the rover, and an early version of the rover's radio received and transmitted data. By the summer of 1993, a year after MFEX had gotten under way, the first version of the SDM was operational. Limited though this first testbed was, it would allow the software guys to try out their motor control algorithms, driving the rover around the building. Most of the time they would just keep the vehicle up on a stand, its wheels hanging in the air, so they could test their ability to command each of the motors without the rover wandering off anywhere.

✳

The control and navigation team was going to be driving the SDM vehicle around. They would need a test area insensitive to the vagaries of the weather, one that looked to the rover like the natural terrain it would someday navigate. Henry cleared out most of the biggest room in the building. Carpenters hammered together a wooden frame about eight inches high and fifteen feet wide by thirty feet long. Forklifts drove through the roll-up door at the front of the building and dumped sand into the frame. When the dust settled, the rover sandbox was ready.

Soon it was time to put a more realistic rover brain onboard. The first hardware that reflected Bolotin's flight design was to be the "Engineering Model"—a printed wiring board version of the rover brain. Like the flight computer to follow, the Engineering Model actually consisted of two boards: the CPU board containing the 80C85, memory, and most of input/output circuitry; and the power board, mostly containing the converters and regulators that powered the many onboard devices.

Within JPL was a group devoted to flight electronics fabrication. Inside Building 103 were cleanrooms and "flow benches" designed to prevent any particles in the air from settling on electronics boards as the flight technicians soldered down components and wires. Their procedures had been worked out over the years to maximize the reliability of flight electronics, since, once launched, repair was impossible. The rover control and navigation team contracted with 103 to turn Gary Bolotin's schematics into the Engineering Model boards.

The weekly reports by the Building 103 engineer coordinating the electronics fabrication fell into a disquieting pattern. The tasks promised for a particular week weren't done, *but we still have plenty of time,* he seemed to say every week. The fabrication engineer said it one too many times. Stone took him off the team.

The engineer was supposed to be ensuring that all of the steps of fabrication were moving along rapidly to guarantee on-time delivery of the final product. But he didn't seem to understand an important rule of flight projects: A day lost now is a day lost forever. Launch dates don't wait. You don't have plenty of time.

Stone took it upon himself to bird-dog the electronics board fabrication. He and Bolotin would together see that the fabrication got back on track. It would take us months to recover from the delays introduced during that dismissed engineer's tenure on the job.

*

Stone thought it would be a good idea to get a countdown clock to remind all of us located in Building 107 of the deadline we were working toward.

He went to George Alahuzos to get it. Alahuzos was an institution within JPL. He had been there since the early sixties, and he knew *everyone.* Stone sometimes called him the Sgt. Bilko of the Laboratory. To me he was expediter par excellence. Over the years, Alahuzos always had his henchmen, one or two skilled technicians, to do the footwork. Now he was teamed with Jim Lloyd. If you needed something pushed through the system, or a way around bureaucratic roadblocks, Alahuzos would see that it got done. If you needed to get a piece of flight hardware made in the machine shop fast, Alahuzos could get your work order slipped to the top of the queue. You just didn't want to dig too deeply into the how of it . . . We were just glad he was on our side. So Stone gave Alahuzos a small budget to find, procure, or build the clock. "I know just what you want. It'll be beautiful," said Alahuzos. Time went by. Alahuzos found an electrical engineer to do some work in his off hours. Stone had thought that the clock might show up in a couple of months, but it didn't happen. Whenever he questioned Alahuzos, the answer was the same: They were still working on it. More time passed. We started to joke that Pathfinder would launch before the clock was ready to announce it.

When Stone finally saw the nearly finished clock, almost a year after he had commissioned it, he was both pleased and appalled. It was huge! It was a rectangular box four feet wide and almost two feet high with a six-inch-tall digital display along the bottom marking the remaining days, hours, minutes, and seconds. The upper half of the box was a sign, lit from behind by fluorescent tubes, that read "MARS PATHFINDER TIME TO LAUNCH." The clock would dominate any wall it was mounted on. And the engineer who designed it had placed a data port on the side so that the time could be programmed via a laptop computer hooked into the Internet to get an accurate time synchronization signal. The clock *was* beautiful. It also reeked of "boondoggle."

Alahuzos's technician mounted the countdown clock on the wall overlooking the rover sandbox. It was a constant reminder of how much there was left to do and how little time remained to do it. A couple of months before launch, Brian Cooper, Henry Stone, and I moved our offices into the second floor of Building 230, the Spaceflight Operations Facility, joining the rest of what was becoming the Pathfinder operations team. We brought the clock with us, mounting it high up on the wall above the cubicles that covered the floor. The clock was big enough to be seen and read from anywhere on the Pathfinder-owned section of the second floor.

After launch, the clock would be reprogrammed and the sign changed to "MARS PATHFINDER TIME TO LANDING."

✳

The installation of components on the Engineering Model electronics boards had just not happened as quickly as needed. When Stone went over to Building 103, he had often found the electronics technicians assigned to his task off working someone else's job. When they were actually doing the work, the flight technicians were excellent. But if a question came up, and they needed to consult with Stone or Bolotin, the board work would come to a halt.

And the supervisors in 103 were telling Stone that his schedule was not realistic. Stone could not accept this. He had deliverables, and he had to find a way to meet them. It was obvious that there was a lot of dead time on his boards. Only if the technicians were working full-time on his job, and still fell behind, would he believe that his planned schedule was infeasible. The problem was that the electronics fabrication group was a service organization, supporting many projects at JPL. There was no particular reason for them to consider Stone's job more important than any other customer's. Every job for them was a flight job; why should they be more committed to Mars than to, say, Saturn?

The experience of building the Engineering Model boards convinced Stone that he needed to find a better way, or he would never succeed in delivering the flight boards on time. To Stone, the solution was to make the assembly people part of his team. He needed to get them moved into Building 107, right next to the control and navigation team's own dedicated electronics technician. That would also put them just downstairs from Stone. If they had questions, there would be somebody right there to answer them. Bringing them into 107 would protect them from the distractions in 103, and ensure that they were working on only one task, the rover electronics boards. Stone began pushing up the line management chain to try and make it happen. The section manager responsible for 103 wanted to be accommodating, but he resisted giving up control of his people, even temporarily. Stone also went to his own section manager.

One day at lunch Stone announced that there was going to be a meeting between him and the two section managers about how to deal with the slipping schedule on the electronics boards. Rather than being pleased, he was worried. Stone never expected much help from line management. For whatever reason, his experience so far on Pathfinder was that intervention by the line organizations tended to slow things down. He figured that the managers would want to avoid stepping into each other's territory, which was exactly what Stone's proposal called for. He expected to get "eaten alive" in the meeting.

Stone was at a crossroads, and he wasn't sure how to handle it. Having the luxury of sitting on the outside of the situation, I offered two pieces of advice. "Whatever you do, don't raise your voice during the meeting. And if you go in there without Bill Layman, you're crazy." I knew he needed someone with sufficient stature at the Lab to back him up. Layman was the only person I could think of. Henry protested that the meeting was only a few hours away, and he didn't know if Layman was available. "Then get on the phone now."

At the meeting in the 103 section manager's office, Stone listed his issues, laid out his case, and argued for moving the rover electronics assembly activity into Building 107. The section manager suggested solutions that would not require such a move. The room was tense. After the discussion had gone around in circles a few times, Layman had his say, as easygoing as you could imagine. He knew how good a group the 103 crowd was, and how excellent their products were. He also knew how good a team Henry was leading. Layman viewed the problems on the table as a breakdown in communication. It didn't really matter where specifically the problem came from; the solution was to get the right people talking to each other. Locating them together in the same place would make that a whole lot easier. The Pathfinder mission and its rover were hugely important to the Laboratory, and we all had to do whatever it took to make it a success. This was bigger than section boundaries, bigger than turf. He was sure that everyone in the room could agree on what had to be done for the future of JPL.

The deal was struck. Stone would have his two flight technicians working in 107.

He had the people. He still needed a facility in 107 where those flight technicians would have the tools to do their work. It was time to upgrade the 107 electronics lab space. And Stone knew how to get it done. He put George Alahuzos on the job. Flow benches, exhaust hoods, and door seals appeared. In a matter of a few weeks, yet another part of the building had been transformed. Building 107 was ready to assemble flight electronics.

※

The first rover that MFEX would build from scratch would be the System Integration Model, or SIM. The SIM vehicle would be identical in almost every way to the final flight rover. The rover team would develop their assembly procedures and refine them from their experience putting together the SIM. If there were mistakes made, they would be made on the SIM, early enough to avoid repeating them on the flight unit. The SIM would do much more than validate assembly procedures. During its voyage in space and its mission on the Martian surface, the flight rover would encounter extreme conditions of temperature, acceleration, and vibration. The SIM would be a guinea pig, forced to undergo a series of environmental qualification tests, experiencing conditions far worse than any we anticipated the flight rover would ever need to endure. The rover team was now engineering a design intended to survive these environments. If the SIM passed through the gauntlet of environmental tests, we would know that the design was correct. But the qualification tests would be harrowing—the SIM would be aged by them. We would never risk aging the flight rover before its voyage had yet begun.

When the time came, the SIM would become the team's "hangar queen," substituting for the flight rover in operations tests as we rehearsed maneuvers her sister would perform millions of miles away.

Second and last off the handmade assembly line would come the Flight Unit Rover, or FUR. When we eventually built and handled the FUR, we would carefully apply all of the lessons learned from the SIM. The FUR would never roll through sandy soil, or even be exposed to the outside world, until that world was Mars. The FUR would be tested just enough to prove that every aspect functioned properly, but not so much that it would experience any appreciable wear to its components or loss to its remaining useful lifetime. It would be pristine.

*

Donna Shirley had decreed that the rover's gender was female. Together with the Planetary Society, a space-exploration advocacy organization headquartered in Pasadena, Shirley also worked out a plan for naming the flight rover, a plan that would involve the public in the mission. The rover would be named after a heroine, real or fictional. Only students would be given the chance to submit candidate names, along with essays describing the key traits of the potential rover namesakes, and how those traits would help the intrepid rover carry out her mission on Mars.

In January 1995, the "Name the Rover" contest was announced in a magazine distributed to science teachers around the country. Teachers told their students, and the entries began flowing in. By the deadline,

3,500 entries had arrived, not only from the United States, but also from Canada, India, Israel, Japan, Mexico, Poland, and Russia.

Selecting the best names and essays now fell into the hands of a small group of volunteers consisting of members of the Pathfinder and rover teams, and Planetary Society staff. Each volunteer read seventy-five or so entries, then joined with the others in the living room of the craftsman-style house on Catalina Avenue that served as the offices of the Planetary Society, to debate the appropriateness of the thousands of possibilities.

When I heard the name "Sojourner," I knew the rover had found its identity. The word meant "traveler," which was exactly what the flight rover was going to be. Sojourner Truth had lived during the mid-1800s, a freed slave who preached for the abolition of slavery and equal rights for women.

The SIM would take the name of the second-place entry, Marie Curie, after the chemist who had discovered radium and polonium in the early years of the twentieth century. Matt Wallace, one of the rover power sub-system engineers, worried over the name Marie Curie. The original Marie Curie, a brilliant scientist, had died of radiation poisoning from the same elements she was famous for discovering. Wallace feared that the name Marie Curie, so long associated with radiation, would draw attention to the Radioisotope Heater Units—RHUs—carried by the flight rover to prevent the WEB electronics from freezing while on Mars.

Nevertheless, both winners were announced to that public on July 14, 1995, the thirtieth anniversary of Mariner Four's flyby of Mars. The SIM had become operational only weeks before. The FUR was still months from completion. But both of them now had names that would go with them no matter how far, how wide, or how long they traveled.

ELEVEN

SEEING AND BELIEVING

As members of the Robotic Vehicles group were being "deputized" out of the research camp into the flight project community, Brian Wilcox picked up the reins of the rover research program. He intended to make the research program as relevant to MFEX as possible. To do so, Wilcox would need a microrover of his own. Fortunately, there was one available: Rocky 3. Dave Miller had departed JPL, and Wilcox's group had inherited the vehicle.

The old Mars Science Microrover had successfully demonstrated CARD navigation of a microrover. Its onboard hazard detection sensors, however, had been virtually useless in sunlight. The new Mars rover would require something better, a "look-ahead" sensor, able to identify obstacles before the rover bumped into them. Wilcox knew that the "machine vision" software his group had developed for the Robby vehicle was not an option. The 80C85 microprocessor on the new flight rover was far too slow to handle the necessary computing.

The flight rover didn't have to have its own path planner onboard: A human operator on the ground would look at stereo images from the Pathfinder lander's camera and designate the rover's path, relying on the same basic Computer-Aided Remote Driving technology Wilcox had first demonstrated almost a decade earlier. The lander's camera was to be called IMP: Imager for Mars Pathfinder. To the project, the IMP was a science payload, not an engineering system. Several science teams had proposed competing designs for Pathfinder's imaging system. Most of the designs had been for one-eyed cameras, without stereo capability. Fortunately for the rover team, the project had chosen the IMP, which produced just the kind of images CARD required to function. The IMP Principal Investigator, responsible for the camera's development and delivery, was Peter Smith of the University of Arizona.

Using Smith's IMP camera, Earth-bound engineers would be able to plan paths for the rover that avoided obvious hazards. But, as long experience had proven repeatedly, dead reckoning error would cause the rover to drift off course during its traverses, bringing it face-to-face with obstacles its human masters had carefully worked to avoid. And there might be hazards that the Earth-based operators would miss, especially as the rover forayed farther and farther from the lander and the IMP camera.

To protect itself from unexpected hazards, the rover needed a reliable look-ahead sensor. Here was a well-defined need where the remaining rover researchers could directly help out the flight project. Wilcox had studied the various navigation sensing approaches researchers had used over the years for robot navigation. He began to consider variations on these approaches that made sense for a microrover. Wilcox knew that whatever he came up with would have to satisfy the typical flight rover requirements: minimal mass, low power, small volume, limited computation.

He settled on a variant of an approach that had been implemented by both robotics researchers and manufacturing industries again and again: "structured light." The idea behind structured light is that if you put a known pattern of light down onto a surface, the bending of that pattern tells you a lot about the shape of the surface. With the right pattern, and a camera to see it, the rover would be able to determine the presence of rocks, drop-offs, and steep slopes directly ahead.

Wilcox's scheme required cameras on the rover.

✳

For a while, it looked like the rover would have no camera at all. Bill Layman remembered: "We'd been wrestling our way through budget cutting and schedule cutting and we reduced scope drastically several times trying to get people to believe that we could build our rover for reasonable cost. I think at that point Donna was ready to abandon anything . . . I mean she talked a lot during that period of time about how the rover's just going to have to run around and bump into things—figure out where it's going by bumping into things," like a blind man without a cane.

Layman continued. "Everybody we talked to that could provide us a camera wanted more weight, power, complexity, and schedule than the whole rover had, just for the camera itself." Wilcox: "The gauntlet had been thrown down that we needed a camera and that it would cost hundreds of thousands of dollars, if not millions of dollars, to put a camera on here, big and heavy and have its own box, and its own power supply, its own processor, and an interface of some form." Wilcox saw that going to the camera development experts at JPL was not going to work. They just had no experience delivering flight hardware for the pittance MFEX had available.

Without a camera on the rover, Wilcox's hazard detection concept would fall apart. And having no ability to detect hazards could easily prove disastrous for the mission. Wilcox had another idea—for a camera of his own—but it was clear that there would be no support from Shirley for developing a camera: She had already been convinced any camera would be too expensive. So Wilcox went to Layman. Layman later summarized Wilcox's proposal and their subsequent discussions: "Brian Wilcox came forward and said, 'Gee, why don't you just let the rover's brain be the television camera, and we'll mode-switch, and it'll sit and think like a television for a while. It shouldn't take too many components.'"

The light-sensitive sensor that "sees" the image is an electronic chip called a Charge-Coupled Device, or CCD for short. Every digital camera contains a CCD. The CCD is arranged as a large array of pixels, corresponding to the pixels you would see on a television screen. The camera lens in front of the CCD focuses an image on its surface. Particles of light strike the surface of the CCD, and cause electrons to build up in the corresponding pixel. The brighter the light hitting a particular pixel, the faster electrons will accumulate at that location in the CCD array. The electrons on the CCD form a pattern of electrical charges that corresponds to the image focused on the face of the CCD. The CCD is the heart of an electronic camera, but it is still only a part. In addition to the lenses that must focus the image on the CCD, there are also electronics to read out the built-up charges associated with each pixel. Once all the charges have been read, the charge must be flushed out, leaving a clean slate, ready for the next image to be recorded.

For an actual TV camera, this process of reading out and flushing the pixels must happen very rapidly, usually thirty times per second. A typical CCD chip might have about 350,000 pixels, so the other electronics that make up the camera ends up reading out over 10 million pixel values each second. The rover's 80C85 microprocessor wasn't fast enough to do that. But we didn't need "live video" from the rover; there would never be enough communications from Mars to Earth to send so much information. All we wanted from the rover was one or two images per Martian day. And for once, the hostile, frigid environment of Mars would help. As Wilcox explained it, "My basic concept was to control the CCD directly from the CPU. Because it was cold you could clock the images in slowly; the charge would just sit out there on the CCD and you would just clock them in slowly, and gather the images as fast as you could, which turned out to be fifty-three seconds for a full image." Using the rover's brain to read out the CCD would also be convenient for hazard detection. Wilcox's approach to obstacle finding would require only a small part of an entire image, a few horizontal rows of pixels, also called "scanlines." Wilcox: "When you wanted to get range data you would flush the scanlines out of the vertical transport registers, so you got one or more rows you wanted to analyze, so you shipped out only those rows. So in a few seconds you could get a few selected rows out of the image and do processing on them."

Cameras onboard the rover would have other uses as well. Wilcox was convinced they would be critical for both scientific and public interest purposes. A rover imager would capture unique close-up photographs of selected targets—rock formations and Martian dirt—that would remain forever distant from the lander's IMP camera. And as the mission progressed, the public would want to see the rover's eye view of its latest travels.

Layman wanted a camera on the rover almost as much as Wilcox. He okayed spending "a few tens of K" dollars on investigating the rover-brain-camera idea. Wilcox drafted Jack Morrison, a software and electronics engineer in his group. Morrison didn't know much about CCDs, yet: "Brian had the concept of taking a bare CCD, and minimal electronics, and interfacing that to the rover computer. So I quickly learned all I could on how CCDs work, and then designed a little circuit that we could prototype to interface to that A/D board, and got that to work. There was a long period there that started out with just trying to get the thing to work, and then getting better and better images all the time, learning how to operate it properly. The electronics that we ended up designing became the basis for what we put on the rover."

About the same time that Morrison was designing his circuit, Donna Shirley asked Brian Wilcox to meet with the JPL camera experts and explain his rover-brain-camera. After all, there was no reason to ignore JPL's existing experience base in flight camera design. And Shirley knew that Wilcox was an expert in robotics, not in cameras; he might be able to use some pointers.

At the meeting, the overall reaction to Wilcox's plan was "It won't work." The main objection was to the many seconds it would take for the rover brain to read out an image. Wilcox's concept basically used the CCD as a storage device for the image. The longer that image remained on the CCD, the more visual noise would creep into it, stray electrons building up and washing out the image. After enough time had passed, the noise would overwhelm the CCD, making it impossible to know what the original image had looked like. The amount of noise was proportional to time, and Wilcox was proposing to read out the image hundreds of times slower than was traditional.

But the noise was also proportional to temperature: the lower the temperature, the slower the buildup of noise. Wilcox was counting on this. Perhaps the camera wouldn't work so well on Earth, but Mars would be colder. Most of the time the Mars air temperature would be below zero, so the impact of noise would be minimal. The others in the meeting remained skeptical.

What happened next was vintage Wilcox. Rather than resort to further technical analysis to convince the skeptics, he proceeded to complete a working demonstration system, and do it on a shoestring. The culmination of Wilcox's and Morrison's efforts came quickly.

Within a week of being told that it wouldn't work, "we had an image," Wilcox said. The picture was a bit noisy, but even at room temperature, you could clearly see the tabletop and equipment in Morrison's office, where the camera sat. And to prove that the camera would work at Mars temperatures, Wilcox and Morrison filled a picnic cooler with dry ice and liquid nitrogen, and stuck the prototype camera in the makeshift freezer. "We cooled it down until it was very cold, then pulled it out just long enough to take an image. Our first little camera was in a pot-metal box with a lens screwed on. It had nothing but a bare CCD and a few bypass capacitors on the voltage lines on the inside of the box. Everything else was on the other side of the cable. The only problem there was ice buildup on the lens. We had to put desiccant inside the box, so that when we cooled it, it wouldn't get ice all over it." The colder the CCD got, the clearer the images got.

Wilcox gave a printout of the first picture to the rover Chief Engineer. "Bill Layman posted the first image on his wall, to tell the naysayers they were barking up the wrong tree . . . We were very pleased to have Bill's support."

They would still need to design flight lenses and custom camera housings to shrink the size and mass. But the rover-brain-camera concept had been proven. The flight rover would have cameras. And those cameras would not break the bank.

*

Wilcox's structured light system would need to be brighter than the Martian sun. The pattern of light produced by the rover would be useful only if it were visible even against sunlit ground. But the sun is very bright, even on Mars, half again as far from the sun as the Earth. How could a tiny power-limited rover generate as much light as the sun?

The secret would be lasers. The sun pours out huge amounts of energy in the form of heat and light. But that light energy is spread over all wavelengths of visible, infrared, and ultraviolet light. A laser channels all its energy into one wavelength. If you are looking for just that one wavelength of light, the laser light can be bright; if you look for any other wavelength, the laser is invisible. Lasers can be tiny, like the type found inside laser pointers. (In fact, for early indoor testing, Wilcox did use laser pointers.) Wilcox went looking for a commercial supplier of small but bright laser diodes. He found one, and found the lasers he needed. The lasers put out light only in a particular wavelength of infrared, making them invisible to the human eye, but very apparent to the CCDs in his new rover-brain-cameras. To make the cameras even more sensitive to the lasers' particular infrared wavelength, Wilcox installed filters over the cameras' lenses, filters that blocked all light except light the same "color" as the lasers generated.

Wilcox drew up a simple optical design to spread the single spot of laser light out into a fan. Once constructed, the "stripe projector" would draw a line of light on the ground. Wilcox made assembling the hazard detection system sound easy: "We got a camera running on the cardcage. We put that on Rocky 3. We got some lasers, built up some optics, and by the summer of '93 had a complete system running with five laser stripers." The final configuration had two cameras along with the five lasers, all mounted together on a rigid camera bar. Just as with stereo vision processing and CARD, careful alignment and calibration would be necessary to a successful system.

The autonomous traverse capability of the rover was activated by the "GO TO WAYPOINT" command. The command told the rover the coordinates of its destination, measured in meters from its starting point. Given these coordinates, the rover headed straight for the target, veering off only if it encountered obstacles along the way. After avoiding any such hazards, the rover would doggedly return to a path aimed at its destination. Once the rover's own estimate of position told it that it had come within about four inches of the target, the navigation software declared victory and the vehicle stopped.

Making "GO TO WAYPOINT" work depended on Wilcox's hazard detection sensor. To check the territory ahead for safe passage, the rover first turned on its lasers, took pictures with its cameras, then shut down the lasers. (Keeping the lasers on only as long as needed conserved power.) If the ground just ahead of Rocky 3 was perfectly flat, the five lasers together formed a symmetric crisscrossing pattern of straight lines on the surface. The stripes of laser light would always be visible in exactly the same place in a rover camera image. But if the ground wasn't flat, the stripes would shift, deformed by the presence of a rock or a ditch. To simplify the job for the rover's slow-thinking CPU, the brain didn't examine the entire image, but only four selected rows of pixels. Examining a single row from an image, a laser stripe would show up as a single spot, the brightest pixel in the row. The amount the spot shifted left or right along the row was proportional to the height of the rock or the depth of the ditch. By putting together the results from five laser stripes and four image rows, the rover created a sketchy topographic map of the terrain just in front of its wheels. The map was just detailed enough for the rover brain to identify whether there was a hazard in view or not, and whether that hazard was a bit to the right, the left, or directly ahead. The rover did all of its hazard checking while sitting still. If the path in front of it was clear, the rover would drive forward a few inches, then stop and take another look.

The hazard detection system had to look out far enough ahead of the vehicle so that when a rock was detected the rover could stop, turn, and drive around it. Wilcox didn't want the rover to be required to back up: The flight design couldn't afford the mass and complexity of putting a second hazard detection system on the rear end. Driving blind backwards would be dangerous. The only way to avoid backing up was to turn in place until there were no obstacles ahead of the rover, and then drive forward in the direction the vehicle was now facing. But how could the rover know it was even safe to turn around without hitting something in the process? The laser stripe projectors had to look out to the sides in front of the rover, not just straight ahead. This made the rover a cautious nearsighted creature. If it encountered two rocks, one on either side, leaving a path along which the rover might pass, the vehicle would still often go around. The rover could only see about three feet ahead, and if the rocks were too closely spaced to allow the vehicle to turn around while between them, there was a chance the path led into a box canyon that could be escaped only by backing up. Rather than risk it, the rover went looking for another way.

This necessary feature of the rover's navigation algorithm was often frustrating to observers, even those of us who knew exactly why the rover was acting in such a timid fashion. From the comparatively omniscient viewpoint of a human being standing nearby in the sandbox, the correct path to the rover's destination was obvious. But the rover's point of view was more that of an infant crawling along the floor, and we would not be able to provide the rover with the benefit of our human perspective while it rolled through the dust of Mars.

Wilcox and Morrison continued refining the hazard detection and navigation software, running Rocky 3 around in the Building 107 sandbox. They rearranged the rocks periodically to discover any weaknesses in their design that might be masked by operating in the same terrain over and over again. There were still questions to answer. When the rover saw a hazard, how far should it drive away from it until it resumed a direct course for its destination? If the rover swung too wide, a short path could easily grow into a long one. But if the rover turned back toward the goal too soon, it might run into the original hazard again, and be forced to waste time avoiding the same rock a second time. And what if the tallest part of an obstacle happened to be between the points in the rover's topographic map, making it invisible to the vehicle's sensors? They needed to strike a balance in the hazard detection software: The rover could be too bold, always driving over traversable rocks, but sometimes failing to see a real obstacle until its bumpers ran into it or it got stuck; or the rover could be timid, almost never running into an obstacle, at the expense of often running away from something it could readily traverse.

With the hazard detection approach largely proven, the MFEX team accepted it as the baseline for the flight rover. They installed cameras and lasers onto the SDM vehicle as part of its upgrade to Rocky 4.2. Fine-tuning of the rover's navigation system would continue until nearly the end of the MFEX development effort.

＊

There would be only one test of the flight rover's hazard-detection system under Martian conditions—at least before it reached Mars. For this test, Sojourner would join the Pathfinder lander inside JPL's twenty-five-foot-diameter solar/thermal/vacuum chamber, and practice its moves in simulated Mars sunlight.

So that the team could observe the activities inside the chamber after it had been sealed shut, TV cameras had been placed at a few key spots. In the middle of the floor we placed a rock to serve as an obstacle for the rover. We couldn't use a natural rock, since it would have contaminated the flight hardware that we had worked so hard to keep clean. Instead, we had welded the "rock" out of sheet aluminum and painted it green.

At the end of all preparations, the Pathfinder lander sat in the center of the chamber, its petals open as if it had recently landed. The rover stood on its designated petal, ready to drive off. The chamber door was shut and sealed. Following a carefully defined timetable, the environment inside the chamber slowly transformed from Earth to the temperature and pressure of Mars. A huge lamp at the top of the chamber stood in for the Martian sun.

At the appropriate moment on this simulated first day on Mars, the rover test conductor commanded the rover to drive forward. The rover complied, stopping only after its rear wheels had cleared the end of the lander ramp. Now that the rover was down on the chamber floor, we commanded it to activate the hazard detection system and go to a way-point destination that would force the rover to encounter our carefully constructed rock. When the rover saw the rock, it would circle around it, then head on to its original destination. This traverse should prove that the whole system functioned, as designed, in the brightness of Mars sunlight. We had already done rover tests outdoors in Earth sunlight, which was twice as bright. There was no reason it shouldn't work . . .

The rover veered off to the left, long before it could have detected the rock. The rover test conductor aborted the command sequence. He sent a new sequence, correcting for the rover's new position and directing the vehicle to its original target. We watched the video monitor, waiting for the rover to move. Instead of moving forward, it started to turn, as if there was something in its way. The floor around the rover was totally flat. There was nothing nearby for it to see as an obstacle. What was happening? The rover seemed to be scared of its own shadow. It was running from hazards that weren't there.

While the rover team scrambled to figure out what was going on, the lander guys were incredulous. Couldn't the rover drive a couple of yards without getting into trouble? We could read the expressions on their faces: "You don't know where it's going, do you?"

The lander test continued. We resorted to "low-level" motion commands to direct the rover to a safe parking location. As long as the rover kept its eyes closed, it did everything we asked of it without complaint.

When the lander system test ended, the rover team spent a few days running the rover around in the chamber before it had to be cleared out for another project's use. After a day of troubleshooting we had found the source of the rover's odd behavior: tape we had laid down to form a grid pattern on the floor. When we pulled up all of the tape, and turned on the Mars "sun" (and got out of the chamber to avoid a nasty sunburn), the rover navigated in the chamber just fine. The ghost hazards were gone.

The tape didn't seem unusual in any way. But the rover cameras had filters on them, making them sensitive only to the laser stripers' narrow range of infrared light. Unfortunately, the tape happened to be extremely dark—almost black—when viewed in the infrared, so when the rover's laser stripe landed on the tape, almost none of the light was reflected back to the rover's camera. When none of the expected laser light was visible to the rover, the hazard detection software had only one way to interpret the event: "Drop-off Detected." And so the rover had run away from nothing.

Would there be anything on Mars that could cause a similar problem? The planetary geologists working on Pathfinder were confident that we would find nothing as dark as the tape we had inadvertently used. But if our landing site proved darker than anyone expected, we'd now be ready for it.

TWELVE

TWO SPACECRAFT

Despite the bias of the rover team, the primary purpose of MESUR Pathfinder was not to place a rover on Mars. Instead, the mission was to prove we could get a payload to the surface cheaply and reliably, to show that the MESUR program's armada of sixteen landers could be accomplished for the estimated budget. And since the intent of MESUR was to create a global network of science stations, it required a lander that could handle a variety of terrains.

But MESUR's days were numbered. Even a $1 billion price tag for an entire network of landers on Mars began to seem too big to NASA management. MESUR was cancelled.

The rechristened Mars Pathfinder went on. The low-cost landing approach being implemented could potentially be employed for other future Mars missions, MESUR-like or not. Furthermore, science instruments could be mounted on the lander, so the first lander mission to Mars in over twenty years would not only prove the landing system, but also provide new data to the science community. Matt Golombek had been appointed Project Scientist. He would see to it that—within the confines of this "technology demonstration" mission—useful science was done.

In the history of the space program prior to Pathfinder, all U.S. landers destined for the Moon or Mars had relied on "powered descent" to the surface, and looked a lot like the Apollo lunar module, with one or more rocket nozzles underneath and landing legs upon which to stand. When the Vikings of the mid-1970s reached the vicinity of Mars, they followed the same procedure the Apollo spacecraft had when reaching the Moon years earlier, looping around the planet and firing onboard thrusters at the right moment to slow into orbit. Later, each lander would again fire thrusters to begin its descent to the surface. Mars has an atmosphere (unlike the Moon) that Viking could take advantage of: a blunt heatshield protected the lander while slowing the rate of fall through the atmosphere to a more manageable speed, until a parachute could be released to slow it further. Then the heatshield, its task complete, would be jettisoned. In the final moments before reaching the surface, the lander rockets would be fired to slow the descent until, ideally, the lander's vertical velocity reached zero just as the legs made contact.

The design worked, but it had its limitations. If the lander came down in a rock field, it might land on top of a rock taller than its legs, causing it to topple over or damaging its payload. You could solve the problem by building the lander arbitrarily big, but you'd pay for it in mass and volume, which would rapidly make any mission prohibitively expensive.

Pathfinder was going to do something different.

The Pathfinder lander was to be a tetrahedron, or four-sided pyramid. There were no legs at all, and no controllable rockets to slow its descent. The pyramid would initially be encased in a protective "aeroshell," consisting of a forward heatshield and a rear backshell, which would shield it during the first phases of the descent. Dispensing with powered descent altogether, Pathfinder would come barreling into the Martian atmosphere on a straight shot from Earth. Small trajectory correction maneuvers during its trip would ensure that the spacecraft did not miss Mars, and would aim it right at the selected landing site. Protected by a heatshield, similar to Viking's, the spacecraft would decelerate from nearly seventeen thousand miles per hour to about nine hundred in only two minutes. A parachute would then deploy while Pathfinder was still falling supersonically. When the parachute brought the lander's speed down below about 250 miles per hour, the heatshield would drop away, and the lander pyramid would lower itself down a sixty-foot-long cable from the backshell. When onboard radar detected that impact with the surface was imminent, huge airbags would inflate from each face of the pyramid, and small solid rockets on the backshell would ignite, bringing the lander to a standstill, hanging about forty feet in the air. At just that moment, the cable would be cut. The lander, cocooned in airbags, would fall to the ground. And bounce. Many times. When it finally came to rest, the airbags would deflate, again revealing the tetrahedral lander. There was no telling which face of the pyramid would end up flat on the ground, but it didn't matter: The lander was self-righting. As the sides of the pyramid would open, like the petals of a flower spreading in the morning sunlight, their contact with the ground would force the lander into an upright position. With the petals fully open, the lander's solar cells, cameras, and science instruments would be revealed, and the landed mission could begin.

At least that was the plan. Some people thought the scheme sounded crazy. But the Pathfinder spacecraft design team had accepted the responsibility of making it work. And the team knew that it must do something different to land on Mars with a budget about one-fifteenth that of the prior landing, more than twenty years before.

※

The design of the Pathfinder lander both helped and hindered the rover. The rover would travel to Mars while tied down to one of the lander petals. Once the petals opened after landing, the rover would be sitting only inches—the petal's thickness—above the surface. But the lander airbags presented a challenge. They might even be the most dangerous hazard the rover would encounter during its mission. By the time the rover was deployed, the lander would have already deflated and retracted the airbags, leaving loose folds of cloth around the edges of the petals. If the rover drove over airbag material, the airbags might catch in the wheel cleats and wrap themselves around the rover's wheels, ensnaring the rover before it could even begin its mission. "Everybody said, 'How the hell are you going to get off the lander?'" Howard Eisen remembered. "And we said, 'Ramps.'" Something like red carpets. "And we had no idea what we were going to do. When we sold the rover to the lander, we had no idea how we were going to get off."

Over at least the first year, the details of the ramp idea remained in limbo. In various rover documents, the ramps were referred to as "drawbridges," or "red carpets." Somehow, the ramps would unroll before the rover, providing safe passage over the airbags and onto the Martian soil. Whatever the rover team came up with would have to stow in a small space, survive the same hostile environments the rover would face, and weigh very little.

Then one day Eisen heard part of a presentation from an outside company: "Some guys from Astro Aerospace were at JPL making a sales pitch for some other program. I pulled them aside, and said, 'Hey, I've got this thing, I want to maybe drive this rover off this ramp or something. What do you think you can do?'" Astro had products based on a stem material that "curls up real nice like a tape measure. When you let it go, either by feeding it out or letting it spool out, it goes ahead and curls out, and it becomes a stiff member." Astro had used its stem material before to form antenna booms on spacecraft. When the stem material was rolled up on a reel, it looked like a flat ribbon of metal tape. When reeled out, the sides of the tape curled up, giving the tape a circular cross-section; the stem material now looked like a long tube instead of a ribbon. This shape made it much stiffer than any tape measure. Which was just what we needed if we were going to trust our precious rover to its stability.

The rover team chose to contract out the ramps. Few companies submitted proposals. The job looked risky, the time was short, and there was almost no mass to play with (only about 4.5 pounds). Astro did send in a proposal, which included a videotape showing a prototype ramp in action. An engineer held the stowed ramp in place, then let go; the ramp unrolled smoothly. The prototype had two of Astro's stems with attached tracks for the rover's wheels and cross supports in between to ensure the alignment of the stems. Astro got the contract. "We thought we were buying into a very simple, very straightforward system," Eisen said.

"The design proved to be anything but simple." Astro's previous systems had been designed to deploy from free-flying spacecraft in zero gravity. The ramps would have to unroll in Mars gravity. They would need to be a yard long to clear the airbags. And they would have to support the weight of the rover over that distance.

To make the ramps stronger, Astro needed to use two stems per side, not just one. The only option was to place one stem inside the other. This forced additional careful design to ensure that the two stems slid past each other, rather than binding.

The original stainless steel of the ramp tracks was too slippery for the rover's cleated metal wheels, so the rover had a tendency to slide uncontrollably down the ramp. Astro added textured surfaces to the tracks to give the rover traction.

The final design did not always unroll as planned. The ramp was like a tightly coiled roll of paper in your hand: When you loosened your grip, it didn't unroll, it just became a suddenly looser coil. Eisen: "We had so much energy in this contained area. When the system was released, the energy caused the roll to expand outward. It 'exploded.' It expanded in all directions at the same time." If the ramp did unroll, it often buckled somewhere along the way. The solution was to use Velcro, which acted like sticky tape between the coils of the ramp: The ramp could now unroll, peeling open, without loosening all at once.

But the Velcro made the performance of the ramp temperature-sensitive. And we wouldn't necessarily know in advance what time of the Martian day we would be releasing the ramps. "We had to deploy anywhere from very early in the morning to very late at night. That range of temperatures was 80°C [145°F]," Eisen said. So they tested the ramp over the full temperature range. Nylon Velcro didn't work at all temperatures: It could either be so stiff that the ramp didn't work at all, or so loose that the ramp acted up as if it were missing Velcro altogether. A special flight-qualified stainless steel Velcro would work, but wore out after only a few uses, which meant you couldn't test it much once you installed it. The testers were running out of options. They tried making one side nylon Velcro and the other side stainless Velcro. It worked great.

When Astro delivered a prototype of the final flight ramp to JPL, the rover mechanical team ran a rover model down the ramp in several configurations. These tests revealed one final problem: If the far end of the ramp wasn't in firm contact with the ground when the rover drove from the lander petal onto the ramp, the ramp could twist, rolling the rover right off and flipping it over. By this time Astro had had enough: The job had proven much more challenging than they had expected. Astro felt that they had delivered on their promises. They wanted to deliver the flight ramps and be done with it. But the JPL team thought they had the answer to the flip-over threat, one that Astro could implement easily: Modify the ramp to have a weak point near the top. Then the ramp would bend at the weak point, tilting the rest of the ramp down until the end reached the ground. The ramp would "break" the way a metal tape measure would when extended out too far. Once the end of the ramp was on the ground, it would be stabilized and the rover could safely drive down.

At first the company balked at the proposed modification. Eisen told them, "You know, it would be pretty embarrassing if we got all the way to Mars, everything worked, we started driving off the end and the rover flipped over. If you want we can put a little label on the bottom of the rover that says, 'This view brought to you by our good friends at Astro.'" Eisen described what happened next: "The very next day they had their best engineer down at JPL. They took the prototype ramp that we were working with and made the modifications on the prototype to do the sort of thing that we were talking about. And we proceeded to drive our model over the ramp, and it worked very well.

"A few days later the flight ramps were modified."

✳

"Minimize the impact of the rover on the lander." This decree was one of the top two or three requirements imposed on MFEX. There were many reasons to simplify the interface between the two spacecraft. The more the rover depended on the lander, the more demands the rover team would have to place on the Pathfinder team, and the more coordination between the two teams would be necessary. Every interface would have to be tested. And that would take time.

All electrical interfaces between the lander and rover would pass through a "separation connector." The lander and rover sides of the connector would either separate when the rover stood up on the lander's petal after Mars landing, or when a pyrotechnically activated cable cutter sliced through all of the wires when commanded by the lander.

The motivation for the electrical connections was primarily to keep the rover healthy while onboard the lander, and to enable the lander to command several events necessary to deploy the rover after landing.

Since the rover would be dormant—totally shut down—for virtually the entire seven-month trip to Mars, it would be up to the lander to keep the rover from freezing or frying. Early predictions by the Pathfinder project indicated that the trip would be a cold one. So it seemed necessary to put a lander-powered heater inside the rover's WEB. That required at least two wires. If the lander was operating the heater, it needed a temperature indicator inside the rover to tell it when to turn the heater on or off. More wires. A lander-powered switch to wake up the rover would require yet more wires, as would pyrotechnic devices to deploy the rover's communications antenna and APXS sensor mechanisms once on Mars.

A big connector added precious mass, and raised questions of reliability. With Bill Layman's urging, the rover team began finding ways to remove wires from the separation connector as the rover's overall design matured.

The new lander thermal models showed that, during the flight to Mars, the lander would be warmer than at first thought. The heater and temperature sensor could be eliminated from the design. The pyro release devices and their wires could be mounted on the lander's petal instead of the rover, so more wires were dropped.

The rover mechanical team suggested a "reed-relay" to make it possible for the lander to power on the rover without any physical connection. If they mounted the relay on the underside of the rover's WEB, it could be activated by a small electromagnet permanently affixed to the lander, mounted just underneath the relay on the lander petal. When the lander powered on the electromagnet for a few seconds, the relay on the rover would flip over and close the circuit, allowing the rover's own batteries to power up the rover's CPU. As the rover brain booted up, one of its first acts would be to flip on the "computer-controlled power switch." This switch was in parallel with the reed-relay switch. Now, when the lander shut off the electromagnet and power stopped flowing through the reed-relay switch, the rover would stay on until its job was done.

In January of 1994, the separation connector disappeared completely. The best interface was no interface at all.

※

The challenges of making the lander work drew NASA's attention away from the rover. Pathfinder's official mission objectives put little emphasis on the rover, which was, in reality, just a payload.

The rover team saw things this way: We were building a spacecraft. The rover had its own subsystems corresponding to each subsystem on the Pathfinder lander: power, attitude control, telemetry handling, thermal control, telecommunications, and even propulsion (albeit very slow propulsion). The fact that neither the project nor NASA headquarters viewed MFEX this way actually worked to our advantage.

It was not until April 22, 1996, barely four months before Sojourner shipped to the Kennedy Space Center, that NASA took note. Perhaps it was because it was becoming clear that Pathfinder was actually going to fly. The impossibly cheap mission to Mars might be a success. Perhaps it was because NASA management was beginning to realize that the public might be paying attention when the experimental rover payload took its first drive. A failure of Sojourner would be an embarrassing spectacle. And unlike most high-profile missions, no one at NASA had been monitoring the rover's development to know whether the damned thing was going to work!

So on that day in April there was another review. This time the board members were from NASA centers around the country. We needed to show convincingly that we had designed and built a mission-worthy piece of hardware. In the end the board members were surprisingly satisfied with the state of the rover. They still wanted to know why so much of the rover was "single string." Why hadn't we built more redundancy into the hardware? Were they judging Sojourner by the standard of the far more expensive missions of the past? My thought was that if we'd had more mass, volume, and money three years before, we would have built a more reliable rover. We had built the best rover possible with the resources we were given.

✸

Every space mission has an emblem. These are embroidered or printed designs that symbolize the mission and its goals, and tie together its participants. Pathfinder had its own. But MFEX was an independent entity. We interacted with Pathfinder, but we weren't Pathfinder; it seemed only fitting to the rover team that we have our own patch that marked our separate identity.

Early on, Howard Eisen had designed a patch for his mechanical, thermal, and mobility subsystem. He proposed a variation of this, with a three-quarter view of the rover against a triangular background, as the official Sojourner patch. The design was simple, clearly represented the rover, and was noncontroversial. It was accepted.

Some of the rover team, particularly the systems guys and the control and navigation subsystem, wanted a more personal design. We all knew we were working on something special. This was a once-in-a-lifetime project. We wanted to come up with the best mission patch anyone had ever seen. And the only people who would get one would be those who had truly worked on the rover.

For months, on and off, the patch was a topic of conversation during the lunch gatherings of rover engineers in the JPL cafeteria. In contrast to the official patch, we wanted something that made the rover appear more like a marauding monster truck, rolling over anything in its path, rather than the microwave oven–sized vehicle we were actually constructing. Art Thompson said he knew a guy who was a professional animator.

A few months later Thompson showed up at the lunch table with his friend's first sketch. The rover looked mean. It looked huge. It was so mean that its front right wheel had clearly just crushed a Martian: About all you could see of the poor creature were its still-struggling arms and its bugged-out eyes. The rover's hazard detection lasers, in reality invisible and carefully aligned, were instead shooting every which way like death rays. Everyone got a kick out of the drawing. Someone suggested adding a Martian mother pushing a baby carriage, madly fleeing the six-wheeled invader from Earth.

I could already imagine our first day on Mars. We would be in Mission Control. The press would be there too. A news cameraman would pan over and zoom in on the rover team patch embroidered on our jackets. The Director of JPL and the NASA Administrator would see the image as it was transmitted to millions of people around the world.

I suggested we leave out the Martians.

Months passed again. The design evolved further. The words "Mars Pathfinder Microrover Team" now surrounded the rover. The patch got bigger. The first set of patches was ready only a few months before the Pathfinder launch, more than a year after we had started talking about it. The patch was the largest, most complex embroidered design any of us had ever seen. The embroidery company didn't even want the job of making them: Each patch would tie up an automated sewing machine for four hours. Thompson and I collected money to get a full set of the patches made. They were pricey. At first a lot of people chose not to buy one: "It's just too expensive." Later they would see one or two rover team members in the halls, wearing the patch on a jacket: "Is it too late? Can I still get one? Gotta have it." Some team members never put the patch on a jacket. They just framed it and put it on the wall, or put it in a safe place to give to their kids someday.

THIRTEEN

TRIAL BY CENTRIFUGE

"The SIM's been destroyed."

"What?" I was dumbfounded. I had just come into work that morning, and the first thing Brian Wilcox told me left me wondering what to do next. The SIM rover—Marie Curie—had been undergoing centrifuge testing, and as of yesterday everything had been fine. What had gone wrong? Wilcox told me more: He wasn't really sure that Marie Curie had been completely destroyed, but he had the impression that it had either been irreparably damaged, or would be out of commission for so many months that it might as well have been. What he did know for sure was that something had come loose at around sixty gravities, and he reasoned, quite logically, that if that had happened, there wasn't much hope that anything had survived.

I told him to keep his speculations to himself, and went off to find out what was really going on.

Centrifuge testing was a means of subjecting the rover to the same kinds of accelerations we expected it to experience during launch and landing. When the Delta II rocket lifted off its pad, and again at Mars arrival, the Pathfinder spacecraft and the rover it carried would feel many times normal gravity. The testing had been going on at Wyle Labs in El Segundo, down by the airport. Most of Wyle's test facilities were located out-of-doors, often covered by an awning that barely protected equipment from the elements. But in Southern California, those elements were not so severe, and you could get away with it. To reach the test site you had to walk down several asphalt-covered alleyways. The centrifuge pit was over thirty feet in diameter and about twelve feet deep, with thick concrete walls. The concrete was marred in places where items, presumably from other customers' tests, had broken free of the centrifuge, been flung outward, and slammed into the walls. During preparations for the rover test, some team members would play the game of identifying the locations where the destruction of a piece of hardware had caused a gouge.

The centrifuge itself was T-shaped and about five and a half feet high. The payload to be tested would be mounted on the end of one fifteen-foot arm of the T, while a weight would be slid out and clamped down on the other arm to provide a counterbalance for the test item. As the centrifuge was spun up, the object on the end of the arm would be subjected to higher and higher acceleration, mimicking flight conditions. No one wanted (or would be allowed) to be in the pit once the centrifuge started spinning: If you pressed yourself up against the concrete wall, the end of the rotating arm might just miss you. At top speed, the arm would be swinging past more than twice per second.

The source of power for the centrifuge was what looked like an old diesel schoolbus engine in an alcove on one side of the pit. The horizontal drive shaft connected the diesel engine and the base of the centrifuge. The drive shaft was painted like a barber pole so it would be obvious to any observer when it began to spin. When it was running, the engine was loud and diesel fumes filled the pit.

This did not look like a place for flight hardware. The rover itself would never be directly exposed to the dirty outdoor environment of the centrifuge pit. There was a cleanroom facility next door. It was there that Marie Curie had been locked down to an aluminum plate in exactly the same manner that it would be stowed on the Pathfinder lander petal. One-eighth-inch steel cables held the rover to the plate in three separate locations. Restraining hooks held each wheel in place, so that the rockers and bogies could not shift. A final restraint held the APXS sensor head tightly against its cradle. Once all of these tie-downs were installed, a clear Plexiglas box was placed over the rover and itself fastened to the plate. The entire assembly was then moved out of the cleanroom and carried downstairs and past the diesel engine to the centrifuge. The box with the rover in it could then be mounted in any of several orientations on the end of the arm.

This was to be a six-axis test. Since the forces the rover would need to withstand during the mission could come from any direction, the rover must be tested in all six representative orientations: wheels down (as if the rover were sitting on a table), wheels up (upside down), front-end down, front-end up (sitting on its tail), left-side down, and right-side down.

As with the other environmental tests, the plan was to subject Marie Curie to "qualification" test levels, while Sojourner would be exposed to "flight acceptance" levels only. "Qual" levels were much more severe than actual flight conditions; the idea was to prove the design of the rover was sound, and would withstand the actual conditions with plenty of margin. It was also understood that qual testing would eat into the total lifetime of the rover. If a qual-level test were continued indefinitely, there was a good chance that something would break. When Sojourner followed Marie Curie onto the centrifuge in a week's time, it would be subjected to only half the total acceleration. This level would verify the workmanship on the flight unit, without reducing its effective remaining lifetime. The combination of tests on Marie Curie and Sojourner would give us the confidence that the unit that actually flew to Mars was ready for the rigors it would face.

After days of setup, Eisen and his team started centrifuge runs with the Marie Curie rover on Tuesday, November 7, 1995. It was just over thirteen months to launch. They had time to test the rover in two orientations before calling it a day: +x, as if the rover was experiencing extreme gravity with its nose pointed at the ground, and –x, with its nose pointed up at the sky. Marie Curie did just fine: It held together the way it had been designed to. The next day they went through the remaining four test cases.

The final axis to be tested was with Marie Curie effectively hanging upside down. This meant that the rover was in a solar array–out configuration, with the wheels toward the center of the centrifuge. During this test run, as with the others, the centrifuge would be run up to the qualification level of 66g's, or sixty-six times the force of gravity on Earth. To achieve this level of acceleration, the centrifuge would need to be revolving at nearly 130 revolutions per minute. This run was to be the last of the day, and in fact was intended to be the last of Marie Curie's centrifuge tests. When Allen Sirota asked for status in the late afternoon, the test team told him by phone that they would be done soon, and he could safely state in his daily status report that the Marie Curie centrifuge test was complete. But during the run, at around 52g's, something let go. The test team didn't know right away exactly what had happened, but they were sure something had moved that wasn't supposed to. When the rover assembly was removed from the centrifuge after the test, it was obvious what was wrong. The front left wheel had somehow come free of its "cowcatcher" restraint and slammed into the underside of the solar array! The steering actuator was actually embedded into the panel.

Several of the solar cells on the array had been cracked by the collision between parts of the rover. And when they checked the state of the batteries, they discovered that for some unknown reason, Marie Curie was powered on. How had that happened? Had the reed-relay switch, designed to let the lander turn on the rover during the long cruise through space and immediately after landing, somehow "bounced" during the test, allowing power to flow just long enough for the rover to boot itself up?

But the rover team had a more immediate concern: Just how badly had Marie Curie been hurt? A cursory external inspection would not answer that question. Electronics, power, and telecommunications subsystems all had to be checked out. The structural integrity of the WEB and the running gear was also in question. The team at Wyle packed up the rover and returned it to JPL by the end of the next day. The rigorous evaluation of the state of the rover would begin Friday morning.

In the meantime, Sirota and others struggled to develop a recovery plan in response to this blow to the always success-oriented schedule. They scrubbed the Sojourner centrifuge test. They would have to fall back to using the SDM for the next set of system tests with the lander, barely a month away. And, assuming that Marie Curie's electronics were undamaged, they would use those electronics in APXS noise testing, even if Marie Curie were partially disassembled for refurbishment. Work would continue on Sojourner.

❋

The engineers in the cleanroom were snapping at each other. "Lookyloos" were showing up to see the wreckage. But the Assembly, Test, and Launch Operations (ATLO) team members were having none of that: Anyone there who was not doing something obviously important was being told to leave or justify their presence. The ATLO team did not want any extra personnel near the wounded Marie Curie. They didn't yet know how bad the damage was, and they were very aware that until they did, the future of the entire rover mission was hanging in the balance.

When Jake Matijevic first saw the rover after its return from Wyle, "I thought it was toast." It wasn't that the rover looked so bad. The only obvious signs of damage were cracks in some of the solar cells. But what about the inside? What might have happened to the wiring, the CPU board, the sensors, and the integrity of the WEB itself? Where was he going to get the money to make Marie Curie functional again?

The inspection of Marie Curie began.

Some of the engineers argued that it would be too dangerous to even power up the rover. They were afraid that when the wheel had come free under high acceleration, its cleats had rammed into the cabling that had been routed along the underside of the solar panel, puncturing and damaging the cables. If that were so, and if unintentional shorts had been created between power- and data-carrying wires, then turning on the rover could cause currents to flow where they were not designed to, potentially burning out many of the components on the electronics boards. But after a half hour of visual inspection, no signs of cable damage had been seen. Continuity tests were performed on individual pins in the connectors on the ends of the cable to see if there were any shorts between wires. None were found.

There were only those cracked cells on the solar array.

During the assessment that followed, the solar panel was removed and the inside of the WEB was examined. Rover software engineers Tam

Nguyen and Jack Morrison ran the low-level software diagnostics. The search for further damage turned up nothing.

We had been lucky.

By the end of the day on the following Monday, the sense of disaster had dissipated. Sirota wrote in his daily email report to the rover team that "Today we embarked on the road to recovery, and we are healing quite rapidly . . . The SIM vehicle should be functional again by Wednesday."

A dozen solar cells on the panel had been cracked in the accident. Surprisingly, the change in power output of the strings of cells was almost unnoticeable, with a maximum degradation of 5 percent in one of the strings.

Within a week of the Marie Curie centrifuge failure, the plan to do Sojourner centrifuge testing had been reinstated. The team was confident that the same failure could not occur at the 33g level of the Sojourner test. While it was easy to miss the fundamental point in the scramble to recover from the damage to Marie Curie, the centrifuge test had served its intended purpose: It was always preferable to encounter a problem during test, rather than in flight, when no correction was feasible.

As we relaxed and proceeded to repair Marie Curie and get back on track, the rumors circulated throughout JPL that the rover had been destroyed in testing. People I ran into would ask about it, wondering what I was going to do now, as if the mission were over. I would tell them that everything was okay, but they seemed to think I was trying to put a good face on a bad situation. I had too much to do to worry about the rumors.

While we implemented the Marie Curie recovery plan, one discovery made after the centrifuge test failure, but initially low on the list of concerns, began to increase in prominence: Marie Curie had gone through the centrifuge test fully powered up! The rover just did not want to stay shut down. The control and navigation guys could not come up with an explanation for what was going on. Nothing in the electronic design could be found that would cause a spontaneous wakeup. And nothing in the software seemed likely either. The problem was a serious one: If the rover could unexpectedly turn itself on, then it might do so during its seven-month trip to Mars, arriving with dead batteries. Without batteries, we could never operate the APXS instrument at night, and any inadvertent traverse by the rover into a shadow would shut it down.

We were now rushing toward a December 15 on-time delivery of the Sojourner rover to the lander. We wanted Sojourner to be as flight-ready as it could be, although we knew that some unfinished rework items would remain. But once we handed Sojourner over to the lander team, it would be mounted on the lander, and would travel along as a hitchhiker during lander environmental testing. We would not be able to get our hands on Sojourner again for many months. So we functionally tested Sojourner as much as possible in the limited time. Morrison and Nguyen calibrated the hazard detection system, navigated the vehicle to waypoints, demonstrated rover stand-ups, and tried out every possible rover command at least once.

The assembly and test of the rovers had become a complex dance as boards were removed, modified, and reinstalled, sensors were calibrated, and the teams split their time between two vehicles in different states of readiness. Sirota's daily "microschedules" orchestrated the flow of activities out of which the microrovers matured.

In early December the ATLO team took Sojourner back to Wyle for a toned-down version of the centrifuge tests that had harmed its sister only a month earlier. This time, the rover survived its tests, apparently unscathed.

✷

The day after Sojourner had been returned from centrifuge testing, Jack Morrison ran through a series of checkouts in the cleanroom to make sure that the rover was none the worse for wear. He found a problem with the APXS Deployment Mechanism, or ADM. The ADM was a simple robot arm on the back of the rover, powered by a single motor. Once on Mars, the rover could place the APXS sensor against rocks or soil by extending the arm out and down. Morrison reported to Sirota that the deployment mechanism would only go one way. You could deploy the APXS, but the motor would not operate to retract it. Sirota's reaction to the news was unexpected: He smiled. The description of the symptom told Sirota exactly what had failed. The only way for the deployment mechanism to move in just one direction was if there had been a FET failure in the "H-bridge" circuit controlling the ADM motor. There were eleven H-bridges on the rover's power electronics boards, one for each motor. These circuits used FETs (Field-Effect Transistors) to switch the power for the motors on and off, and control their direction. So what had caused the FET to fail?

At the next meeting of the ATLO team, Sirota mentioned the trouble with the ADM. Howard Eisen expressed surprise. "It worked just fine when we ran the motor down at Wyle." Something triggered in Sirota's head. What Eisen had said was somehow important. Wyle Laboratories was the place where they had done the centrifuge testing. The mechanical team had done most of the setup for those tests, which required stowing the rover and locking it down to the aluminum plate that simulated the lander petal. Locking down the rover required driving each wheel independently to get just the right tension on the wheel cages. The APXS Deployment Mechanism also had to be operated to put the ADM in its simulated flight configuration so that it could be tied down to the plate, just the way it would be on the way to Mars. The mechanical team had built Ground Support Equipment to power each of the motors in turn, without having to power up the rover brain itself. The GSE was a box with ten toggle switches on it. There was a switch for each steering motor and each drive motor, so you could run them forward or backward at will. Ten toggle switches. What about the eleventh motor, the ADM? Sirota remembered that he had seen the GSE to operate each of the motor drives, but he had never come across equipment to drive the ADM. The connectors for the ten mobility motors were easy to get to, located on the front of the WEB. But the connector for the ADM motor was on the back of the WEB, and the ADM itself tended to get in the way. It was suspicious that the only motor giving them problems was the same one that, to his knowledge, had no GSE box to run it. How did they operate the ADM motor?

Eisen told him that he would just connect the output from a power supply directly to the motor leads themselves by poking sharp probes through the coating material, and the motor ran just fine.

The pieces of evidence in Sirota's mind came together. It clicked. Sirota asked the next question, but he knew the answer. "Did you remove the ADM connector from the rear bulkhead before applying power to it?"

Eisen's mind was working too, somehow perceiving a threat. "No. Why?"

"Well, if you put power on those wires, it doesn't just go out to the motor. It also goes through the connector onto the rover power board, right back to the H-bridge. The FET isn't built to take the large reverse bias you put on it every time you applied power to the connector." So of course, eventually, it had failed.

The lights were going on around the table. The engineers on Eisen's team understood the explanation. And they believed it.

And the power didn't stop at the FET. The power fed right into the rover's power bus, and woke up the rover's CPU. Even though Eisen was feeding power into the rover for only the few seconds it took to deploy the ADM onto the plate, that was enough time for the CPU to boot up and do what it always did, bring battery strings B and C on-line. Once that happened, the rover could stay on until it was commanded to shut down again, or its batteries went dead.

The following Monday they opened up Sojourner to take a look. The FETs for the ADM motor were on the top side of the top board in the WEB. There it was: a blackened spot on the board surrounding the FET, where it had overloaded, overheated, and burned the coating material covering it. But once again, the rover team was lucky, for there was no other damage to the board.

Once he understood what he had done, Eisen was chagrined. Most of the rover team was just relieved that they finally understood the mechanism by which the phantom wakeups had occurred. Mysteries that did not yield to reason and analysis were disconcerting. Knowing that human error could fully account for the observed evidence was reassuring to the engineers.

It *was* satisfying to the rest of the team to see Eisen publicly admit a mistake. In the prior two years, Eisen had lambasted others for their errors, or for acts that Eisen had decreed were errors. He was often called the "Teflon engineer" because blame never stuck to him: He was too skill-ful at deflecting it. He was very smart, an exceptionally capable mechani-cal engineer, and he was driven to be right about everything.

Immediately after the mysterious wakeup incident, Eisen's contrite-ness made him easier to work with. Had he learned a lesson? No. The pe-riod of his humility was short-lived, only a few weeks. Eisen was like a warrior momentarily stunned by a blow. Soon he shook it off and was fully recovered, ready to do battle again. Eisen was a key member of the rover team: We needed both his aggressiveness and expertise to get the job done, so we put up with the day-to-day frictions.

✳

As we pushed toward the goal of handing the Sojourner rover over to the lander, the delivery date receded from us. The lander simply was not ready to receive the rover into its integration and test activities. The date for the handover began to slip day by day. The rover team used the addi-tional time to attempt corrections to an ongoing noise problem with the APXS. Then, in the early morning of January 23, 1996, the carefully boxed Sojourner was moved the few hundred feet to the Pathfinder cleanroom in the building next door, where the lander folks had been working long shifts to keep their own assembly, integration, and test operations on track. Despite their efforts, the lander team was twenty days behind schedule. Somehow, they had to get those twenty days back. On February 1, the Pathfinder project held a review of the rover documentation and status, and officially accepted delivery of Sojourner. With that acceptance, Sirota announced the end of the rover implementation phase, and the start of integrated spacecraft testing and rover flight readiness prepara-tion. It was a distinction drawn in the sand: Old rover issues remained un-resolved, and the first threat to integrating Sojourner with the lander had already revealed itself.

FOURTEEN

CAN WE TALK?

I walked through the card-key protected doors into the Pathfinder test control area. Computer workstations, assigned names of characters out of *Star Trek* for easy identification, were everywhere. Engineers were wearing jeans and headsets with push-to-talk microphones. Some of them munched on breakfast while studying their screens.

At the far end of the room a long, low window looked out into another world. The high-bay cleanroom of the Spacecraft Assembly Facility was a huge room, with a ceiling at least forty-five feet high. A uniform river of air flowed constantly across that room from one end to the other, sweeping up suspended particles too small to be seen. High-efficiency filters cleansed the air on its way to another transit of the high bay. The atmospheric pressure inside the high bay was slightly higher than outside, so any breach of the room's seals would cause clean air to leak out, and no dirty air to seep in. You entered the high bay only by airlock. While trapped in the airlock, you would be subjected to a high-pressure "air shower" that removed any loose dirt you might be carrying with you.

The partially assembled Pathfinder lander rested on a mobile platform a few yards from the window through which I peered. Beyond the lander I saw the spacecraft cruise stage and backshell. Each one had been mounted in its own aluminum framework, which could be rotated and tilted for easy access by the lander assembly and integration team. The high bay had been built to accommodate much larger spacecraft. The merging elements of Pathfinder were dwarfed in that volume.

The engineers on the other side of the thick glass were dressed far more formally, not in suits and ties, but in aptly named "bunny suits": lightweight white smocks and pants, white hairnetlike caps over their heads and white booties over their shoes.

Human beings are dirty. They are shedding all the time: hair scales, flakes of skin, moisture. Particles they've been carrying around on their clothes with them. An average person easily generates over 2 million particles per minute. The bunny suits were there to protect Pathfinder from its creators, cutting the shedding to only a few tens of thousands. The carefully designed air flow of the high bay carried away most of the dust that remained. Partly, the intent was to minimize biological contamination: This hardware was going to Mars, and we didn't want to contaminate any potential Martian life with microbes from Earth. NASA's planetary protection policy would not allow Pathfinder to launch unless the spacecraft was clean. Just as important, we didn't want any particles settling inside a fuel line or on a circuit board. A thruster that failed to fire, or an electronic component that shorted out, might end the mission before it started.

Humans also carry electric charges, static electricity which can destroy electronic components. The fabric of the bunny suits contained a grid of conductive thread to dissipate such charges. Conductive straps ran from the wearer's foot to the underside of the white booties, to contact the conductive floor of the high bay, preventing charge buildup.

Also in the high bay was Sojourner. It was January 23, 1996. The rover had been delivered just hours before. Now it was time for the "Surface Operations Mode" test, the first joint test of the flight rover and the Pathfinder lander. This test would be the first chance to practice some of the interactions of the two spacecraft planned for the first day on Mars. The flight lander would send command sequences to Sojourner, which would respond with telemetry data. For weeks prior, I had worked with lander engineers to define the detailed procedure we would follow today. Every activity involving the Pathfinder flight hardware required an approved, signed-off procedure. This was the only lander we had: The procedures were our first defense against human error during the myriad of activities between now and having a functioning spacecraft on its way to Mars.

Art Thompson and other rover team members, wearing their own bunny suits, watched over Sojourner in the high bay. I remained in the test control area, ready to send commands through the lander to the rover. We were separated by less than twenty feet. Yet the only way to communicate with Thompson was either via the communal headset voicenet, or like visitors and inmates in prison, staring at each other through the glass and talking by telephone.

The test conductor slowly stepped us through the procedure. Confirm the test-cable connections to the lander. Check the voltage levels on the power supplies. Apply power to the lander bus. Finally, the lander was up and running. The rover telecommunications guys had previously delivered and installed the lander-mounted rover radio and antenna. One of the engineers at a computer workstation sent the lander command "MODEM_POWER_B" to turn on the lander-mounted rover radio; the next command was "ROVER_WAKEUP." The rover, fresh out of its delivery box, was powered up. We waited. But no data flowed across the rover-lander radio link. With no commands from the lander, the rover made no moves. The lander and rover generated only error messages.

The test had failed. Or rather, it had never even begun. Lander-rover communications wasn't functioning.

The team dropped into debugging mode. Almost immediately, the test conductor wanted to skip the rover test and move on to the next procedure.

The rover-lander link should have worked. Art Thompson on the rover side and Glenn Reeves on the lander side had already consumed weeks in the testbed, testing and debugging communications.

The link should be working!

✳

The lander and rover teams had each spent so much time in the past proving out their pieces of the radio interface that each had the same reaction:

"It must be *your* problem. What's wrong with your side of the interface again?"

"By this point, we were all from Missouri," Thompson said. "We wanted the other guy to show us that his interface was working correctly before we'd consider that there might be a problem on our side." At the end of the two days allocated to the surface operations test, the lander and rover were still not speaking. The lander integration schedule was tight. The engineers working the communications problem would have to retreat to the testbed to continue debugging, while spacecraft integration moved on to the next critical activity.

The communications failure looked suspiciously like a problem Thompson and Reeves had isolated in the testbed months before. At the time, they had traced the problem to a design flaw in a chip on the lander electronics board. The bug had been a big deal in September. But it had been fixed! All of the computer boards had been sent back to the manufacturer and the bad chips had been replaced with new ones that didn't have the flaw. Why would the old symptoms suddenly appear again now?

Within three days of the failure, the lander team had checked their records and confirmed that the computer board installed on the lander had never been shipped back to the contractor to get the bad chip replaced. Every other board, including those in the testbed and the final flight computer board intended to eventually replace the one now installed, already had the fix.

The good news was that we understood the problem. The bad news was that there was, for now, no way to correct it. Opening up the lander to swap out the bad board was not an option: It would take a full week to disassemble the lander enough to get to the board. The lander team was already weeks behind schedule. The soonest they would have a chance to replace the bad board would be mid-May, almost four months away. And even that date was in question, since one of the easiest ways to make up schedule would be to delay installing the final flight computer board until just a few weeks before shipping the lander off to the Kennedy Space Center.

※

This was not the first difficulty with the rover communications subsystem. And as with many other elements of the rover's design, Donna Shirley and her team had broken the rules to get this far.

Most spacecraft communications systems were designed to send signals over long distances. If a spacecraft were in orbit around the Earth, those distances would be measured in hundreds or thousands of miles. If the spacecraft were traveling to another planet or through deep space, the distances would be millions or even billions of miles. But Sojourner would never need to communicate with the Earth, at least not without help. The rover's radio wouldn't have to be heard across interplanetary space; it just had to talk to the Pathfinder lander, at most only a few hundred yards distant. The lander would be Sojourner's communications relay. The lander would have more of everything to do the job: more power from its larger solar arrays, a "High-Gain" antenna that could be pointed at the Earth, and a comparatively huge computer memory to store its own data and that of the rover.

From the start of the MFEX effort, Shirley had decreed that the rover's radios should be a purchased commercial product.

This was not how JPL did things. For each spacecraft, the JPL communications section would study the mission requirements, determine the necessary specifications, then either design and build the communications system in-house, or contract it to an appropriate company experienced with flight hardware. Either approach would likely cost several million dollars.

Shirley knew her $25 million total rover budget was stretched tight. She didn't think she could afford a huge chunk of that budget for designing and building a custom flight-qualified communications system. Lots of companies produced radios, and even radio modems that allowed two computers to pass digital data between them. And from the standpoint of talking to each other, the lander and the rover were just two computers. Radio modems on Earth could converse over miles of ground. There must be an existing commercial radio that could meet the rover's needs as is, or slightly modified. Shirley preferred spending her limited money on the robotic aspects of the rover that made it unique.

Even as the Pathfinder Project Manager was attempting to pressure

Shirley into using a tether instead of any radio at all, a communications engineer on Shirley's team was doing an industry survey to locate commercial radios that might one day serve as Sojourner's link to its operators. Two rover radios would fly to Mars, one mounted on the rover, and its twin installed on the lander. By the late summer of 1992, the best candidate had been identified: the Motorola RNET 9600 radio modem.

Once the Motorola radio was chosen, several were ordered, and a couple of those were sent down to Building 107. As the first Software Development Model rover took shape on the gutted chassis of Rocky 4, one Motorola modem was part of it. The second radio was wired to the cardcage that simulated the lander side of the communications system. For the lifetime of the rover development effort, this radio link would be the preferred means of sending commands and telemetry.

The JPL communications engineer who had selected the radio moved on, leaving JPL for a new job. Soon thereafter Lin Sukamto came onboard the rover team as the communications Cognizant Engineer. It was her first flight project assignment. The job of Sukamto's telecommunications team now became one of repackaging the Motorola modems for flight, then proving that they would survive the rigors of the flight environment: vibration, radiation, and temperature. They would also have to design the rover's antenna and its mate on the lander.

The radios were only a few hundred dollars apiece. They were so cheap that Sukamto ordered thirty of them. Then Sukamto and her team began a methodical, rigorous screening process to select the "best of the lot."

※

For two radios to communicate, the receiving radio has to be tuned to the same frequency as the transmitting radio. Radios depend on tiny components called crystal oscillators to regulate their frequencies. Quartz-movement watches owe their accuracy to their own quartz crystal oscillators. The crystals vibrate millions of times per second. As long as the environment around the crystal remains fairly stable, the crystal oscillates almost exactly the same number of times from one second to the next. This allows radios to consistently transmit and receive at the same frequency. Each rover radio had two crystals, one to control the frequency of its transmitter, one to control the frequency of its receiver. But the crystals could not maintain their frequencies under all conditions. As the temperature of each crystal went up, so did its frequency.

Testing showed that if the temperatures of the rover and lander radios were close, the communications link worked fine. But if the temperatures drifted far enough apart, the frequencies could shift to the point that the radios could lose the ability to talk to each other. The Motorola modems were just not intended for the temperature extremes of the Martian environment. The custom-built radio systems on JPL spacecraft readily handled wide temperature ranges. Frequency drift would not have been a problem if the radios had been designed and built in-house. But in that case, of course, the cost itself would have been the problem.

By late 1995, the rover communications team had still not come up with a complete solution to the frequency drift issue. As Bill Layman would have said, they had not "killed the problem." They turned to Jim Parkyn, one of the JPL communications technical gurus. He proposed a solution: install temperature-compensated crystal oscillators.

One option was a specially packaged crystal, encased in its own tiny "oven." The oven kept the crystal toasty warm at a fixed temperature. Whatever happened to the temperature outside the package, the crystal inside stayed constant, and so did its frequency. But you paid a price for this feature: The component was bigger, drew extra power for the oven, and would not fit inside the existing radio. The rover didn't have the power to spare, so Parkyn suggested other crystals packaged with additional circuitry that corrected for changing temperature. These temperature-compensated oscillators still wouldn't fit into the radio, but at least they were low power and available commercially.

It was far too late to incorporate the change into the original rover design. The question was whether the flight rover could be retrofitted in time. Sojourner was already mated with Pathfinder, undergoing system testing. We would have to implement the fix, install it on Marie Curie, and prove to ourselves that we would be able to safely modify Sojourner in the short time the flight rover would be in our hands again before delivery to Florida.

We bought oscillators. Scot Stride, one of the rover telecommunications engineers, remembered examining the oscillators when they arrived. "They were huge! They were enormous! Putting two of them side-by-side, it's almost as big as one of the radio boards. And we had to build a board for both of those." The inside of the rover was already tightly packed. Finding space for another electronics board with the new oscillators, mounting the board safely in that space, and wiring from the board to the existing radio modem was going to be a challenge. With the rest of the rover already designed and built, there was always the fear that new changes might create more problems than they solved. Scot worked with the rest of the team to test the effectiveness of the new components. "Electrically we got one side to work. It worked really well." The radio's frequency stayed locked in over wide variations in modem temperature. "Mechanically, it was a nightmare." Attempts to integrate the oscillators continued over several months, as the sole opportunity to rework the guts of Sojourner rapidly approached. But a reliable solution eluded us. "Trying to implement it on one of those radios would have been really messy. And we just ran out of time."

Sojourner would fly as is.

＊

For months after the bad communications chip had been discovered on the lander computer board, the rover team used clever workarounds to keep Sojourner involved in the lander system testing. The eventual resolution of the problem was anticlimactic. The computer board that would actually fly to Mars was finally installed on the lander in late May. The next system test confirmed that Sojourner and Pathfinder communicated as they were designed to do. With each successful communications session, the rover team could relax a little more.

＊

Jan Tarsala was a JPL communications engineer.

One day, six months or so before the Pathfinder launch, he ran into Jim Parkyn in the JPL cafeteria. Tarsala and Parkyn were friends. Besides being engineers in the JPL Telecommunications section, they were both amateur radio operators. It was through their mutual avocation that they had first met. Neither of them completely left communications engineering behind when they went home. They had radios in their blood.

Parkyn was cheerful. "I've got some news for you. I've got a job in industry, and I'm leaving the Laboratory." Tarsala knew that Parkyn had been looking for some time. "Did you know that Lin van Nieuwstadt is also leaving?" This was news to Tarsala, but not a surprise. Lin Sukamto had recently married, her husband was Dutch, and they were relocating to Europe. Parkyn went on. "You know what all this means. You are going to end up with the Sojourner radio job."

That was a surprise. "What are you talking about?" Tarsala was already up to his elbows with the radios on another space mission. He protested that he couldn't possibly take on another task.

Parkyn persisted. "No. No. Lin is leaving. You are the heir apparent." It would only be a matter of time before section management came knocking on Tarsala's door. "They are going to come to you, and they are going to say, 'This is your job.' And you are not going to be in any position to turn it away. Let me tell you what you need to know about this job."

So they sat down in the cafeteria, and in about a half hour Parkyn gave Tarsala the technical history of the Sojourner radios. Parkyn warned Tarsala: "You're going to land on Mars, and you're going to be off-frequency. And you are going to have to come up with a way of managing that situation and making it work." Parkyn described the attempt to add the temperature-compensated crystal oscillators to the radios. These would have solved the frequency-drift problem in hardware. He had pushed as hard as he could to get the new oscillators implemented, but somehow Lin and the rover team had shipped the hardware without the fix. "Just know what you're getting into. You're not going to be able to refuse it. Be fully prepared for what's going to happen."

Tarsala kept what Parkyn told him to himself. "I trusted Parkyn. I trusted his technical observations. I trusted his opinion. Parkyn knew the practical side of Radio with a big R."

Sure enough, the Telecommunications section manager came to Tarsala a week or so after Lin van Nieuwstadt left for Europe. "I've got a job for you. You will be working for Sami Asmar," who would be the new Cognizant Engineer for the Sojourner radios. "The whole section is committed to having a successful mission." The section manager turned to Tarsala, pointed at his chest, and said, "But I expect you to be responsible for making this radio work." As Parkyn had forecast, the job was not offered to Tarsala, but commanded. And thanks to Parkyn, he had been forewarned of the challenges that were yet to come.

The target

Courtesy of NASA/USGS

The Blue Rover

Courtesy of NASA/JPL/Caltech

The pantograph: Don Bickler's first high-mobility vehicle design

Courtesy of NASA/JPL/Caltech

Robby: the first excursion into the Arroyo Seco

Courtesy of NASA/JPL/Caltech

Tooth: a tabletop rover

Courtesy of NASA/JPL/Caltech

Go-For: the fork-wheeled microrover

Courtesy of NASA/JPL/Caltech

Rocky 4: the Mars Science Microrover

Courtesy of NASA/JPL/Caltech

The photograph from the Viking 1 lander I showed in my presentation to new employees: What might be over that horizon?

Courtesy of NASA/JPL/Caltech

Sojourner on the benchtop

Courtesy of NASA/JPL/Caltech

Sojourner in the twenty-five-foot chamber with the Pathfinder lander for the final thermal/vacuum test *Courtesy of NASA/JPL/Caltech*

The unofficial rover team patch

Courtesy of Calvin Patton

Sojourner team members in the JPL MarsYard with Marie Curie. *Top row, left to right*: Firenze Pavlics, Hank Moore, Tam Nguyen, Dutch Sebring, Matt Wallace, Lee Sword, Fotios Deligiannis, Ron Banes, Howard Eisen, Ken Jewett, Henry Stone, Jim Parkyn, Art Thompson, Jack Morrison, Allen Sirota. *Bottom row*: Brian Cooper, Jake Matijevic, Lin Sukamto (van Nieuwstadt), Beverly St. Ange, Fred Nabor, Andrew Mishkin, Scot Stride. *Courtesy of NASA/JPL/Caltech*

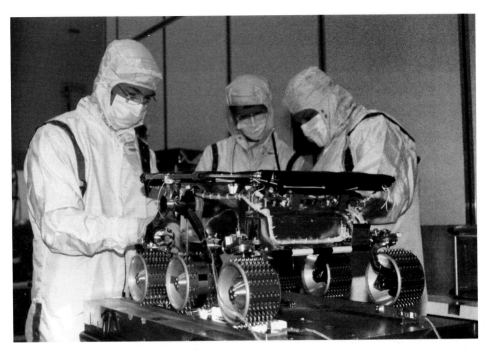

Sojourner checkout at Kennedy Space Center *Courtesy of NASA*

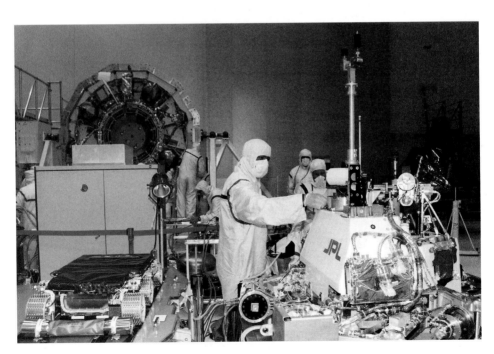

Sojourner joins Pathfinder at the Cape *Courtesy of NASA*

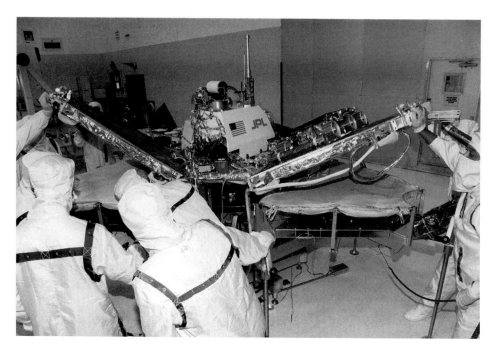

Closing up the lander for the last time *Courtesy of NASA*

Pathfinder and Sojourner
on their way to Mars

Courtesy of NASA

Landing Day: Sojourner and Pathfinder on Mars *Courtesy of NASA/JPL/Caltech*

Sojourner's first images from Mars: views of the forward ramp before and after deployment. Some data was lost due to the rover-lander communication problem. *Courtesy of NASA/JPL/Caltech*

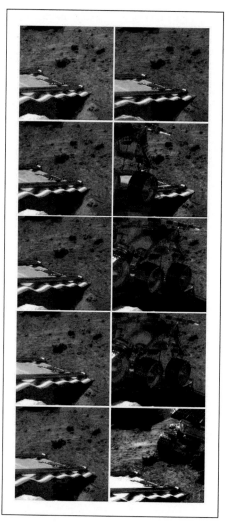

The first rover movie: six wheels on soil (The yet-unseen rover causes the ramp to shift in the early frames.)

Courtesy of NASA/JPL/Caltech

Sojourner touches down on the Martian surface.

Sojourner hits Barnacle Bill on the first attempt.

Sol 3: the first image of the lander taken by Sojourner *Courtesy of NASA/JPL/Caltech*

Sol 5: the second rover image of the lander *Courtesy of NASA/JPL/Caltech*

Sol 8: Sojourner takes a picture of one of its hazard-detecting laser stripes, proving it can be seen in Martian daylight. *Courtesy of NASA/JPL/Caltech*

Sojourner bags the rock called "Yogi," as seen from the lander.

A rover-eye view of Yogi

During a soil experiment, the rover does a wheelie,
lifting its front left wheel into the air.

Courtesy of NASA/JPL/Caltech

Sol 35: Sojourner sits near the rock "Wedge." Part of the Rock Garden is visible in the upper right of the image. *Courtesy of NASA/JPL/Caltech*

Sojourner spies sand dunes behind the Rock Garden. *Courtesy of NASA/JPL/Caltech*

Two generations: Marie Curie in the cleanroom with the twin Mars Exploration rovers

FIFTEEN

THE NOISE THAT WOULDN'T DIE

Allen Sirota's father had been an electrical engineer at Ford Aerospace in New York. Eventually he moved the family out to California, and got a job at North American Aviation, working on everything from the X15 rocket plane to the Apollo moon program. Sirota remembered growing up watching Walter Cronkite on TV describing the Mercury, Gemini, and Apollo missions. When he got into UCLA, he tried several majors but—inevitably it seemed—ended up in electrical engineering.

After ten years in the aerospace industry Sirota realized that he wasn't really enjoying the job he found himself doing. He had always been drawn to the idea of working at JPL, but had never viewed it as a serious possibility. "I thought you needed a Ph.D. to sweep the floors here!" Sirota later said in his JPL office.

When he finally went in to interview at JPL, he was asked about a number of job qualifications. "Sure. I can do that" was always Sirota's response.

He was hired. Sirota reviewed what he had promised. "Can I really do that?" he asked himself. Well, he would find out. He was immediately assigned to be the technical manager on a small flight project, an experiment that would fly on the Space Shuttle. When he arrived, the project was already behind schedule, over budget, and in need of a "miracle." He took these problems as a challenge. Sirota was indeed up to the task, and later looked back at this first JPL assignment as his favorite project of all time. He was too new to the organization to be keyed into the workplace politics and he was given free reign to solve the technical problems. Never again would he enjoy such blissful ignorance.

But working on a rover going to Mars wasn't half-bad. And the experience he had gained learning to deal with Principal Investigators on smaller flight experiments was about to be useful.

✷

With the exception of the lander's IMP camera, the rover's APXS was the key science instrument on the Pathfinder mission because it would determine what things were made of. When the instrument was placed against rocks, it could, given enough time, produce spectral data from which the elemental composition of the target material could be discerned. By knowing the types and relative abundances of atoms (such as iron, sulfur, and so on) in a rock, you could learn a lot about what kind of rock it was.

The APXS was in two parts: the sensor head, which was outside of the rover, to be placed against targets; and the electronics box, mounted inside the WEB, which over several hours compiled the raw data streaming from the sensor head into a complete picture of the target. The APXS depended on nine small pieces of curium mounted in the sensor head. These bits of radioactive material emitted alpha particles that would strike a target rock whenever the APXS was deployed against it. Some of the alpha particles bounced back from the target. Sometimes protons in the target material were also dislodged, and sometimes the alpha particles excited the target's atoms to produce X rays. Detectors in the sensor head picked up the pattern of alpha, proton, and X-ray radiation reflected back from the target. By analyzing the pattern, you could identify the elements in the sample, and their relative abundance.

The APXS was to be a joint effort of the University of Chicago and the Max Planck Chemical Institute in Mainz, Germany. Rudi Rieder, from Max Planck, was the APXS Principal Investigator. This made him the lead scientist responsible for developing and delivering the instrument to JPL for inclusion on the Pathfinder mission. Rieder's counterpart at the University of Chicago was co-Investigator Thanasis "Tom" Economou. Economou's style was naturally combative. When he wasn't making demands on us, he was arguing with Rieder. Rieder was much more easygoing, up to a point. Together, they were an ongoing challenge to the rover team. The interaction between the JPL team and the APXS developers was a classic clash of cultures: flight hardware engineers versus university scientists.

In late February 1995, Rudi Rieder arrived in Chicago, having traveled from Germany with the APXS electronics destined for installation inside Marie Curie. From O'Hare Airport he took a cab to the University of Chicago, where Tom Economou waited. Within days they would be putting the instrument through qualification testing at JPL, subjecting it to the thermal and vibration environments it would see in space.

On the way to the university, Rieder asked the cabdriver to stop at a Starbuck's coffeehouse at the corner of South Harper Avenue and East Fifty-Third. While the taxi waited outside with the engine running, Rudi went in to grab a coffee. At that moment, a fifteen-year-old fleeing police slammed his car into the rear of the cab. The taxi driver went to the hospital, but the APXS electronics in the trunk were undamaged. Rieder smiled for the photographers and told the press: "We don't need to do a shock test. It's been done in Chicago."

The Marie Curie vehicle began taking shape in the middle of 1995. Sojourner followed soon behind. By late October, Allen Sirota was collecting test spectra with the APXS electrically wired to the naked Sojourner CPU board on a benchtop. Economou did not like the results: The spectra were full of electrical noise, masking their meaning. They were supposed to show as graphs with several sharp peaks, like a silhouette of a mountain range against the sky. The positions of the peaks would represent the different elements in the material being analyzed. Instead, the spectra were muddy; all of the peaks that should have been there were blurred into one, as if you were looking at the mountains through smoked glass. Noise was masking the APXS signal. If the noise got bad enough, the APXS instrument would be useless, unable to distinguish the elements in a rock.

Electrical noise was leaking into the APXS from somewhere in the rover electronics. A likely culprit was the power supply that provided the electricity to operate the APXS electronics. If this were the cause, then capacitors placed across the wires supplying the power should filter out the noise. The team implemented this minor modification. But when they took new spectra, the noise was only slightly improved. Economou's reaction to the problem was typical, and destined to become his favorite refrain: "The Russians gave us good power. Why can't you?" There were four APXS instruments included in the payload for the Russian Mars '96 mission, scheduled for launch at about the same time as Pathfinder. Economou insisted that the noise problem was unique to the rover.

If the power supply was not the source of the noise, then perhaps the cable connecting the APXS electronics and the sensor head might be acting like an antenna, picking up the noise radiated from some other part of the rover. But where exactly was the source of the noise?

To find out more, we'd have to put Sojourner together.

✳

On November 1, 1995, Allen Sirota sent an email report to the team: "I would like to announce the birth of the flight rover Sojourner Truth, as noted by the first application of power in an integrated configuration. Early test results indicate no problems at this time, although much more testing lies ahead. To protect Sojourner from any possible harm or contamination, the following guidelines will be adhered to when working on or in the proximity of the vehicle . . ."

✳

By mid-November, APXS testing on the flight rover could resume. Matt Wallace, from the rover power subsystem team, led the testing. The spectra were as noisy as before.

As the rover team continued debugging the problem, Wallace found himself working only with the abrasive Tom Economou. We wondered where the instrument Principal Investigator was. Rudi Rieder might be the needed buffer between Economou and the rover team, and might have some ideas the other scientist didn't. When we asked for Rieder's help, we were informed that he was in Russia, and was unavailable. The JPL engineers felt like they were under the gun to get this problem fixed; yet they couldn't get the full resources of the APXS team to help. Everyone assumed that Rieder was busy preparing additional APXS units for the Russian Mars '96 mission. Whatever the reason, we would not see him again for months.

Now the push was on to deliver Sojourner to the lander by December 15, 1995, only a few weeks away. Sirota had created a plan to meet this deadline. There was a lot of environmental and functional testing of Sojourner that had nothing to do with the APXS. Continued testing of the APXS would threaten the delivery schedule. Sirota decreed, "The APXS noise problem will no longer be investigated until after FUR delivery." For the next month the rover team turned its attention to making Sojourner as flight-ready as possible.

The Pathfinder Project Manager had other plans. Tony Spear wanted the APXS noise corrected before accepting the rover delivery. With lander integration running late anyway, Spear offered to extend the rover's deadline to give the team more time to find a solution. So when Sojourner became otherwise ready for delivery on the eleventh of December, we instituted an APXS "tiger team," led by Matt Wallace, to work the problem at an accelerated pace. Spear and Project Scientist Matt Golombek were taking a keen interest in our progress against the noise. They began holding weekly meetings with Economou and representatives of the rover team.

Wallace knew that whenever the APXS was operated by the rover, it produced noisy spectra. He also knew that if he set up the APXS on a benchtop, with its own independent power supply and totally isolated from the rover, it would deliver clean spectra. His plan was to start out in the bench configuration, and then very slowly, step by step, modify it toward the in-the-rover configuration. He would take spectra all along the way, waiting for the first sign of noise. He would do his experiments on the Marie Curie rover, protecting Sojourner from any unnecessary handling. With Economou present, the search for the noise began anew.

Wallace's systematic, deliberate approach quickly achieved some success. The tiger team traced the noise to APXS cables passing too close to the rover power supplies on their way to the APXS electronics. The signals from the APXS sensor head were carried to the APXS electronics box by four coaxial cables. Due to the rover's thermal design, the cables were forced to follow a circuitous route. The sensor head sat on the back of the rover, while the opening of the cable tunnel "igloo," the only way into the WEB, was at the front of the vehicle. So the cables were routed along the side of the rover, across the front, through the labyrinthine path of the cable tunnel, and finally into the WEB proper to connect with the APXS electronics. This made for a long set of cables; the longer the cable, the weaker the signal, and the bigger the chance of picking up stray radiated noise.

The rover engineers rerouted the coax cables inside the rover as far away from the power supplies as possible. To further protect the cables, they were then wrapped in copper shielding.

With the hardware modifications in place on the Marie Curie rover, Economou viewed the spectra. He declared the instrument performance "adequate" to fly. He hoped that we would make further improvements. Economou would not be pinned down. He refused to reveal any quantitative measure of what would be "good enough." Doing so would allow the rover team to declare victory. But that would mean they would stop work on the APXS. Economou preferred to keep the team focused on his instrument: The more time they spent on it, the better the data it might produce. And Economou wanted the best data he could get.

We declared victory anyway. Sojourner was ready for delivery. We had a solution to the APXS noise problem running on Marie Curie. The solution would be implemented on Sojourner in about six months, during the already allotted rework period.

*

When the rover team returned from vacation after the first of the year, the plan had changed. The project had delayed the delivery of Sojourner yet again, and asked for the Marie Curie noise fix to be implemented on Sojourner immediately, before delivery. The rover team agreed.

The rover flight technicians rerouted the cables from the sensor head and added the copper-braid shielding to Sojourner, duplicating exactly what they had done on Marie Curie. So the first spectra collected in the new Sojourner configuration were a shock: The data plots still showed pronounced noise. Dismay was thick in the room. Until that moment, the team thought they had the problem licked.

The debugging of the APXS resumed, going on for so long that it began to feel like a voyage, an odyssey in a strange land from which the rover team might never return.

*

Matt Wallace had volunteered to lead the newest incarnation of the tiger team that was investigating the APXS noise problem. He was a skilled power system engineer. More than that, he had an even temperament to help him deal with the antagonistic Tom Economou. Prior to his career at JPL, Wallace had been in the Navy, serving two tours of duty on nuclear submarines. He'd been given that assignment partly due to his psychological profile. He could handle long confinements in close, claustrophobic environments.

But working with Economou took its toll. It was just not in Economou's nature to consider the JPL engineers as partners in the effort to solve the noise problem. He could see them only as adversaries, either uninterested or actively seeking to thwart his desire for good APXS spectral data.

Experience working with the Marie Curie, Sojourner, and laboratory model APXS units had taught the tiger team that the supposedly identical APXS electronics had subtle differences that gave them varying susceptibility to noise. This variation had contributed to the team's frustration during debugging: They would sometimes find an apparent solution using the laboratory unit, and then discover it was ineffective on Marie Curie or Sojourner. In the final attempt to correct the noise problem before the Sojourner rover delivery, the tiger team tried mixing and matching electronics boards from all three APXS units. Perhaps some combination of boards would produce clean spectra under all conditions.

After they put the hybrid together, Sirota was ready to declare success. There was still some noise in the X-ray spectrum, but the alpha and proton patterns looked "quite good." The team would do a final evaluation of the APXS hybrid in the Copper Room.

The Copper Room was just that: The floor, walls, ceiling, and even the door were lined with copper sheets. The copper formed a Faraday cage, which shielded whatever was placed inside from any electrical noise outside. There was a possibility that the remaining noisy readings from the APXS were due to unknown noise sources in the building where they had done the earlier testing. If that were so, then there might be nothing wrong with the APXS or the rover, but only with the conditions of the test. Rerunning the tests in the Copper Room could prove that.

When Sirota and Wallace examined the spectra in the Copper Room, they were as noisy as before. The X-ray spectrum was still corrupted. The noise source really was somewhere inside the rover. When the spectrum looked bad, Economou was very vocal. "The Russians gave us good power. Why can't you?" Sirota described the moment: "Economou flew off the handle, accusing us, telling us it's our fault, that we have bad power. Matt had just had enough. He walked right up to Economou's face and started yelling at him. He definitely lost it a little bit right there. I remember trying to pull him back a little bit. There was no physical contact . . .

"After that week, Matt Wallace never worked on the APXS again." Sirota became the liaison between Economou and the rest of the rover team. Apparently he was the only engineer who could work with Economou for an extended period of time without wanting to wring the scientist's neck.

Time was up. There was nothing left to do except reassemble Sojourner and prepare the rover to be handed over to the lander. Further efforts to solve the remaining noise problem would be restricted to working with Marie Curie. Any improvements we were able to implement on Marie Curie would be installed on Sojourner when the flight rover was again in our hands, a few weeks before we shipped it to Kennedy Space Center for launch.

Many rover team members wondered why we couldn't just fly the APXS as is. To them, the APXS was just an add-on to the important thing, which was the rover itself. And given Economou's combative attitude, nobody wanted to go out of his or her way to help him out. It seemed to the engineers that Economou was just getting what he deserved for being accusatory, secretive, and demanding. Some of the engineers joked about installing an APXS ejection system on the rover: Once Sojourner was on Mars, we'd send the command to fire the pyros and explosively jettison the APXS sensor head, sending it flying yards away, never to trouble the rover again.

Sirota's attitude was different. He recognized the importance of the APXS. To the project, it was one of the premier science instruments on Pathfinder, and the rover was just an APXS delivery system. Sirota had worked on flight experiments before, and the science Principal Investigators had been king: The engineers had to jump to deal with whatever the scientists asked for. Pathfinder, as a "technology demonstration" mission, was a bit different. The Principal Investigators did not have such despotic authority. But Sirota was still quite comfortable accommodating the needs of the science instruments. And Sirota figured he could get along with anybody, even Tom Economou.

The Pathfinder project remained vitally interested in the quality of the APXS spectra. The troubleshooting effort went on, now coordinated by Sirota. The weekly meetings with the Project Manager continued. During these meetings, Sirota pressed for Economou to provide a concrete definition of when a spectrum would be "good enough." Otherwise he'd be working the APXS problem until launch. Spear and Golombek backed Sirota on the need for a measurable definition of success. Economou relented and gave them a number. "Getting that number was a major accomplishment," Sirota said.

Sirota got his own harsh message out of the meetings: "The APXS noise problem was a go/no-go situation for the rover. I had the strong impression that Spear did not want to fly the rover without a functional APXS." But now Sirota had to contend with a shortage of rovers. Sojourner was tied up in integration with the lander. Marie Curie was overbooked, with every rover subsystem team competing for test time on the vehicle.

The solution was to build a new APXS testbed out of assemblies left over from earlier environmental tests: the qualification rover electronics boards, a test WEB, and the lab unit APXS. At first Sirota wasn't even sure the testbed could be constructed, since the test WEB wasn't quite the same size as the others, and therefore the electronics might not fit. But by early March, the mechanical team had integrated the pieces. Testing could begin once more.

Sirota worked closely with one of Economou's engineers. As they exercised the testbed, it began to look like the team's prior efforts had not been in vain. The capacitors they had installed in October had reduced the low-frequency noise in the power supply lines. The rerouting and shielding of the cables the tiger team had tried in December had eliminated the radiated noise. But there was still a third source of noise to be dealt with. "This afternoon, we began tuning the filter to block higher-frequency noise," Sirota reported by email. "The results we achieved started to come very close to those which we got from the APXS GSE alone, indicating that it is the high-frequency noise components which are causing the [remaining] APXS spectral noise."

Sirota wasn't quite ready to declare success: "I am certainly encouraged by this discovery, however, given the history of this thing I wouldn't be surprised if I am overly premature and what I am telling you turns out to be a bunch of bologna." But when the filter was added to the Marie Curie rover, the spectra were clean.

They had beaten Economou's number. Now all they had to do was prove that the noise filtering would work for Sojourner. Getting to the flight rover was for now impossible: It was mated to the lander, undergoing environmental testing. Months would pass before Sojourner could be pulled out of integration testing for its own APXS fix.

SIXTEEN

SOUL OF SOJOURNER

"**W**hat the hell's the rover doing? It won't communicate. It's not listening to us. What's going on? Do we have a hardware problem?" Henry Stone was exasperated. A hardware problem on Marie Curie would be scary, because it might mean a similar failure was lurking on Sojourner.

Stone and Tam Nguyen had pulled me into the testbed to see if I had any ideas. They had been doing a standard "healthcheck" of the rover, the kind we would do after launch. They were proving that the whole process of sending command sequences through the lander to the rover was working.

And it was. When the rover woke up, it grabbed the commands from the lander, operated the APXS to show that it too was still healthy, and sent back telemetry. Then Marie Curie put itself to sleep, just like it was supposed to.

Stone and Nguyen woke the rover up again to give it some more commands manually. But by the time they were ready to tell the rover what to do, it had stopped asking for commands.

"What mission phase is it in?" was my first question. That much information was available from the debug port before the rover stopped talking.

Stone answered. "Phase 2. What difference does that make?"

"It thinks it's on Mars. It's running its contingency mission."

❋

The rover operated in several mission phases. The mission phases were our way of protecting the rover from doing the wrong thing (like trying to drive around inside the lander if it was still in space on the way to Mars), and, even if the telecommunications subsystem failed, to make sure the rover did something right. There were distinct mission phases active on the rover before launch, while in transit through space, when sitting on the lander before it stood up, and after it had rolled onto the Martian surface.

During normal mission operations, the rover team would send commands to tell the rover which mission phase to switch to. In addition, the rover had its own set of programmed rules that it checked every time it woke up. If the right conditions were met, Sojourner would take it upon itself to progress from the current phase to the next. So if the lander powered-on Sojourner during its long voyage to Mars, the rover's onboard sensors would detect zero-gravity instead of one Earth gravity, and proceed directly to the "cruise" phase. When Sojourner woke up the first time after landing, it would find itself in fair-to-middling gravity (actually, 38 percent of Earth gravity), and move on to its first "on-Mars" mission phase.

And for each of the phases there was a separate "contingency sequence" already built into the rover to tell it what to do if commands from home stopped arriving. Suppose the rover inadvertently woke up in cruise, and the lander had no commands to give it. The contingency sequence would kick in. The rover would still do the right thing: go back to sleep, conserving its batteries. The on-Mars contingency sequences were designed to make the rover do everything from standing up, to driving down the ramp, to blindly trying to find rocks for APXS measurements.

❋

So what was wrong with Marie Curie in the testbed?

Nothing. The robot had performed exactly as it had been designed to do. The rover had started out in phase 0, or "prelaunch." When Stone and

Nguyen ran their first test, they sent the standard, prewritten cruise health-check sequence that we already had on the shelf, ready to use several months from now, after Sojourner was in space. They queued up the sequence on the testbed lander. The lander waited dutifully, ready to transmit the sequence as soon as Marie Curie requested it. Stone powered-up the rover, and a few seconds later the lander and rover were talking to each other. The rover pulled over the healthcheck sequence. It started to march through the commands. Step 1: Switch to cruise phase. Step 2: Do a healthcheck command. Step 3: Send a few sample commands to the APXS. Step 4: Shut down.

The second time they woke up the rover, Marie Curie was already in the cruise phase. But the onboard accelerometers indicated much more than the zero gravity the rover expected during cruise. There was only one place the rover was going after cruise, and that was Mars. So the gravity it was seeing must mean that it had reached its destination. Marie Curie jumped into the on-Mars phase, phase 2. When it asked the lander for commands, there were none; Nguyen hadn't yet loaded the new commands. With no new commands, the rover began executing its contingency sequence. On Mars, a possible cause of faulty communication between the lander and rover would be that the rover's radio was too cold, so the contingency sequence commanded the rover to turn on the radio heater for ten minutes before trying to contact the lander again.

Stone, Nguyen, and I waited a couple more minutes. Marie Curie's silence ended. The testing resumed.

***

Jack Morrison was Sojourner's software "architect." He was a one-in-a-thousand software engineer, seemingly able to do in a week what a merely human programmer might accomplish in a month. Morrison, together with Tam Nguyen, formed the entire rover onboard software team.

Morrison liked to be left alone. He put it this way: "Social activities are not highest on my list." He had worked at JPL in the 'eighties, then he and his wife moved to Colorado where they lived in a house in the mountains on thirteen acres of land. When his employer began having financial difficulties a few years later, Jack and his family returned to Southern California. Brian Wilcox eagerly hired Jack back into the Robotic Vehicles group.

After his return, Morrison was just as private as before. No one knew much about him. He was always affable when approached. But he never said much about his personal life. He never joined the rest of the group in the cafeteria for lunch, preferring a microwaved meal in his office. I guessed that he viewed the lunch hour as "quality time" to get more work done, with fewer interruptions than usual since most everyone else was out of the building.

Morrison's principles for the rover software: "Early on when I started looking at what I thought the requirements were in an overall architecture, I laid down some philosophies about how I thought it ought to be done, what the main goals of designing it would be. It had to be very reliable. It had to be as simple as possible, mainly because of the constraints we had on the CPU and memory and power. And we wanted to have some visibility into what was going on, since it was going to be an engineering experiment, and we didn't know everything about the environment it was going to be operating in. Looking at that, I approached it as a typical embedded system, with a limited computing environment, always having an eye on how much memory something's going to take, how efficiently something's going to work, and how simple and straightforward you can implement it and have it do what you want."

Morrison's philosophy was perfectly in line with the rover team's desire to keep the overall rover design as simple as possible. This was exemplified by the choice of the 80C85 microprocessor as Sojourner's brain. The computer chip was far less capable than the average home computer of the time. But that simple CPU chip controlled a suite of sensors and motors that would be the envy of any personal computer owner.

The rover's apparent simplicity was deceptive: It would often surprise its creators by doing exactly what it had been designed to do. The combination of all of the design choices made by all of the involved engineers had resulted in a final system that, at first, no single designer fully grasped. This complexity gave the rover something like "personality." While the team members were learning the nuances of the system, it sometimes exhibited behaviors that confounded people.

Unlike many other computer-based machines, the rover only did one thing at a time. "We can't walk and chew gum at the same time" was how Henry Stone described it in presentations. Over its radio link, the rover would accept a new command sequence from the lander and store it onboard, then start carrying out the commands one by one. Once a command was finished, the rover would store the results into one or more telemetry packets, and transmit the packets to the lander. Then the rover asked the lander for a new sequence, just in case a new one had arrived that might override the one the rover was now executing. If there was no new sequence—there usually wasn't—the rover moved on to the next command.

✳

Sojourner had first been brought to life in late 1995. The rover had been delivered to the lander less than three months later, with only minimal functional testing. The original schedule showed more time, but the needs of the hardware had taken precedence. Centrifuge and acoustic tests at Wyle Labs, modifications to the rover brain, and the seemingly unending struggle to track down and fix the APXS noise problem fully consumed Sojourner's first few months of existence. Functional test time dwindled to only a couple of weeks.

In his low-key way, Morrison warned that he needed more time testing his software on the vehicle. I complained to the rest of the core team that the software guys needed more time. The hardware guys nodded and kept tight reign over the vehicle. After all, what was the point of testing software if the hardware wasn't ready? They figured the software testing could happen later. They had hardware to deliver. At the expense of lost functional testing, they delivered it.

On May 17, 1996, we would be reminded that the assertion that everything else would take care of itself was merely wishful thinking.

On that day, Sojourner came out of hibernation. During the past few months, the flight rover had been tied down to the lander petal, exactly as it would be during its voyage to Mars. The lander had been closed up into its pyramidal form with Sojourner inside. Together, the two spacecraft had been subjected to various tests that simulated their shared environment during the seven-month cruise phase. During the entire experience, Sojourner had remained dormant, just as it would throughout the real trip to come.

Now the lander sat on a platform in the cleanroom, its petals open again at last. A few bunny-suited engineers stood watch over the proceedings. A set of tables had been pushed together in a cluster in the next room. Computer workstations were arranged on the tables. The test conductor and the rest of the test team worked at their keyboards or watched the activities in the cleanroom via closed-circuit television. Step by step, they marched the lander and rover through a simulation of their first day on Mars.

The rover's NASA sponsor had flown in from Washington, D.C., to observe the test. He and Jake Matijevic stood off to the side, watching and sometimes speaking back and forth in low tones.

The cable-cutter pyro fired and the single rover ramp unrolled across the table. A few of the engineers applauded. More pyros fired, releasing the lander's grip on the rover itself. Those of us in the test control room watched the video monitor. Sojourner began to rise off the petal. Its rear wheels drove forward, while the rest stayed unmoving, still held in place by their restraining hooks. The rover rose. Almost. Almost. It stopped too soon! The rockers dropped back down. The stand-up had failed.

The telemetry transmitted from the rover showed no anomalies. The rover didn't know it had a problem. The sequence was still running. In a few minutes, the rover would try driving out of its remaining restraints. After a quick consultation over the voicenet, we decided to complete the stand-up manually before that happened. In the cleanroom, a white-smocked rover mechanical engineer lifted the rockers until the bent springs between the articulated halves snapped into place. The rover's telemetry still showed no problems.

There was a hint of movement—a barely perceptible roll backwards. Then the rover stopped dead. Now what? The telemetry came in, with the rover reporting an error. Tam Nguyen looked up the obscure error mes-

sage, which led us to the APXS Deployment Mechanism. For some reason, the rover thought the APXS was deployed, so it was refusing to
budge. That made some sense: We had designed in a fail-safe for the deployment mechanism. When the rover got to Mars, we would be using
the ADM to put the APXS sensor head into the dirt, leaving the sensor
there for hours at a time. What would happen if we forgot to retract the
ADM before driving on to the next target? You didn't want to accidentally
drag the sensor head through the dirt. If you did, you'd likely damage the
APXS, or at the least cover it with soil, which would make it useless for
further experiments. So Jack Morrison had built a self-test into the rover
software. Whenever the rover started to move, the software compared the
ADM potentiometer reading to a preset threshold value: If the new reading was bigger than the threshold, the software would conclude that the
ADM was deployed, and rover motion was disallowed.

You could look at the rover and see that the ADM was stowed. The
rover thought otherwise. We had a mystery, one we didn't have time to
ponder if we hoped to continue the test. "Tam, let's tell the rover that the
pot is bad," I suggested. The rover had a way to deal with sensors it knew
were broken: it used an alternative sensor, if there was one, or more often
just ignored the bad sensor. Perhaps if the rover considered the ADM potentiometer broken, it would be willing to move again.

No such luck. When we tried it, we got the same error message again.
I was stumped. The lander testing went on, but the rover was done for
the day.

Our NASA sponsor seemed to be understanding in the face of the
rover problems. He expressed relief that the rover had survived its cruise
testing with the lander.

The rover core team meeting on the following Monday was full of
gloom, doom, and finger-pointing. Howard Eisen called the failure of Sojourner to stand up during the test "a fiasco." Matijevic said it was a "wakeup
call." Suddenly, everyone in the rover team meeting had ideas for "fixing" the
problem with software and operations: There should be troubleshooting
guides written ... The software team should be more dedicated ... More
personnel should be trained to understand the software ... The core team
should begin discussing operations issues during the weekly meetings ...

I fumed. I had been warning the team for the past year of the hazards of skipping software testing. I had always been told, "We understand the risk, but the hardware comes first." I had raised the issue each time the schedule was cut, so much so that no one wanted to hear it anymore. Henry Stone had warned me that the rest of the team was tuning me out. And now most of the team was surprised that there had been a problem? Even through my anger, I feared that the new attention to the issue would not bring solution, but instead a set of naive quick fixes that would merely compound the initial error.

Later, Jack Morrison expressed the software engineers' reaction: "It wasn't a wakeup call to us. It was a wakeup call to them to start paying attention to what we were saying."

✳

Within a couple of days, Morrison had fully analyzed the rover failures. Rover stand-up was a tricky procedure. The rover drove its rear wheels forward until sensors on the rocker-bogies indicated that the rover had stood up just enough to latch the "broken" rockers in place. Then the wheels stopped. If the rear wheels kept driving past this point, they would start slipping, the cleats on the wheels would start to get bent, and the rocker-bogies would be subjected to stresses they would never encounter in normal driving. The precise values of the rocker-bogie sensors that translated into "Stand-Up Complete" could be determined only by putting Sojourner into its stood-up configuration several times to account for variations in the sensor readings. Due to their abbreviated test schedule before Sojourner's delivery in January, Morrison and Nguyen had never completed the sensor calibration. So the rover had just followed its instructions, rising from a reclined position to what its sensors said was full stature. Mission accomplished. Only the rockers hadn't locked into place, so as soon as the wheels stopped driving, the rover sagged, as if it were weak in the knees. This was exactly the failure that contact switches on the rockers would have prevented. Unfortunately, it was too late to add the switches into the design now.

The practical solution to the problem: Calibrate the sensors. And maybe improve that algorithm for standing up the rover, so that if something did go wrong, the rover would detect it and skip any future motions, thus avoiding even worse trouble.

Nguyen and I had almost solved the second failure in those tense minutes during the test. We had correctly interpreted the error message and traced the problem to the APXS deployment mechanism. The ADM position was indeed beyond the fail-safe threshold. But we hadn't known that, even with the ADM potentiometer tagged as "bad," the rover would remember the last known position of the ADM, and refuse to move given the most recent information it trusted. We learned from Morrison that what we should have done was change the threshold to a better number. The threshold value had been improperly set to an impossible value five months before: The number was so small that no position of the ADM would have worked. In the rush to deliver the FUR, there had been no time to review all the values and catch the error. Any attempt to drive Sojourner after the parameter had been set would have revealed the problem.

Ironically, none of these problems were with the rover software itself. Given the calibrations and parameters it had been provided, Sojourner had done perfectly everything it had been asked to do. Only the results had not been perfect.

✳

The acrimony within the rover team went on for about two weeks. Much of it was evidenced in email traffic among the team members.

When I calmed down, I composed a long email message describing my concerns, how we had gotten into the current situation, and what I thought we needed to do to get out of it. Primarily I was making a plea for more test time. The philosophy in the past had been to postpone software testing until "later"; I warned that "later is now."

Eisen responded with a scathing email telling me all the things the software team was doing wrong and should now do differently. He questioned the commitment of the software engineers, while reaffirming that his mechanical team was "prepared to do whatever it takes."

I quickly banged out my answer to what seemed to me a wanton diatribe, then rewrote it again and again until it read as if my temper were under control.

The email volleys died down. The team moved on to other immediate issues.

After all the Sturm und Drang, little changed. Morrison and Nguyen did create an on-line troubleshooting guide to explain error messages. Marie Curie and Sojourner remained overcommitted, their presence required in system tests together with the lander. As Matijevic viewed it, there was "precious little" time available even for these tests. Days originally scheduled for the software team to test the rover and learn how it performed were still often lost to dealing with hardware issues.

Tony Spear insisted that the APXS noise fix, already demonstrated on Marie Curie, be implemented on Sojourner early, instead of waiting until the scheduled time in July. Spear wanted proof that the flight APXS was ready to fly, even if it meant pulling Sojourner out of project system tests. The APXS surgery on Sojourner was a success. Perhaps there would now be at least a little time for testing with the lander, with a bit more left over for testing on our own.

For the next few months it seemed to the software team that they were always either participating in lander system tests, or getting ready for the next one. There was nothing ideal about the conditions, but in all these tests we were getting to know the rover. Slowly, we got better at operating it.

And the rest of the rover team began to appreciate just how complex a beast our little Sojourner really was.

SEVENTEEN

LICENSE TO DRIVE

From the start of MFEX, Donna Shirley was extremely conscious of her limited budget. The cost-capped $25 million for the rover had to cover everything: not just design, parts, development, assembly, and tests, but all operations during the mission as well. Shirley's cost concerns led to her decision to rely on commercial radios for rover communication. She also hoped that the rover control station already developed for the Mars Science Microrover demonstration would be good enough for commanding the new flight rover.

But the flight rover would not be like the Mars Science Microrover. Sojourner would respond to different commands, and there would be many more of them, for operating its cameras, instruments, heaters, and behaviors. And every control station we had ever created had sent only one command at a time: You told the rover to do something, you waited to see the results, then you told it to do something else. That approach had been just fine for research. But when the new rover got to Mars, we would only get a chance to send commands about once per Martian day. So we had to make those chances count. Instead of sending individual commands, we would send a sequence of many commands—maybe hundreds of commands—all at once, and all in the proper order, which would tell the rover what to do over the next twenty-four hours or more.

Nearly a year after the MFEX team had come into being, Shirley finally relented: "Bring Brian Cooper onboard." Stone and I had convinced her that cobbling together a few improvements to the Mars Science Microrover control station would not enable us to operate the Pathfinder rover on Mars. Cooper had worked on the control station for every JPL rover since the days of the Blue Rover. The true complexity of commanding the flight rover would be revealed during MFEX development. In the end, only about 10 percent of the rover's commands would have anything to do with telling the rover where to go, with the rest associated with the rover's onboard instruments and with maintaining the health of this six-wheeled interplanetary spacecraft.

Brian Cooper became a one-man subsystem, single-handedly developing the software for Sojourner's Rover Control Workstation (RCW). Now officially a member of the control and navigation team, Cooper began attending Stone's weekly team meetings, but quickly got bored: "All Henry ever talks about is parts." Cooper knew that ensuring that Sojourner's components arrived on schedule was important, but it wasn't very relevant to designing the RCW. There was one parts procurement Cooper did care about: We ordered a newer, faster Silicon Graphics Inc. (SGI) computer upon which he would create and run the software that would be the RCW.

It took a long time for Cooper to believe that Pathfinder was real. "I figured it would get cancelled." Or maybe Pathfinder would fly, but the rover wouldn't be onboard. People at JPL were always talking about rover missions to Mars. He remembered when Mars Rover Sample Return had been a big deal. Then it had gone away. Now they were saying that the latest microrover was going to Mars, and he was part of that mission, designing the newest vehicle control station based on everything he had learned from all the workstations he had worked on for the past eight years at JPL. He didn't want to set himself up for a big disappointment. He wouldn't permit himself to become too emotionally involved in the mission. He'd go home at night and talk to his wife: "They haven't cancelled it yet. We'll see." But Pathfinder didn't disappear. Its budget didn't get cut. The test model rover evolved into an ever more flightlike configuration. And, in time, parts of the SIM and the FUR began to come into existence.

Cooper began to allow himself to believe. "We're really going to Mars. We're really going to do this." He opened the door partway to view the enormity of what he was a part of. Like many of us on the team, Cooper had grown up reading science fiction. He and I had often discussed the reason working at JPL was special, and as Sojourner took form, it applied more than ever before: "We get to turn science fiction into reality."

※

As Henry Stone's system engineer, I was constantly thinking about the mission the rover would perform on Mars and what we would need to tell it to do. I didn't realize it at first, but my perspective was already that of the eventual operator of the completed rover. It just seemed easy to know what I was going to need from the control station when the time came. And Cooper was excited by the power of the new SGI hardware and software tools that he now had in his hands. Cooper and I started having long talks about what the Rover Control Workstation should look like. He was the software designer and developer; I fell into the role of customer and user. Together we defined the requirements for the RCW's design.

During an "average" day on Mars, the rover would wake up, drive to a new location, dig one wheel into the soil to study its properties, take a picture, move to a new site, turn around, deploy the APXS against a rock, start collecting APXS data, then shut down for the night. Overnight the APXS would continue gathering data; the rover would wake up for a few minutes every few hours to read out the data collected so far, then shut down again. When morning finally came, the rover would wake up and ask the lander for whatever new batch of commands had recently arrived from Earth. Orchestrating these activities would take two or three hundred commands. For those of us operating the rover, it would be like writing a new computer program every day. And like a stage entertainer, we had to hit our marks on time: The rover had to reach its destination for the day before the lander aimed its IMP camera and took pictures of where the rover was supposed to be. Those pictures had to be sent back to Earth before the day's communications window closed. If the rover wasn't in them, we wouldn't know enough to plan the next day's driving.

For planning the rover's traverses through the terrain, we'd put on stereo goggles like those we'd used for the Mars Science Microrover, and switch to the RCW's CARD display. Within this display window, we could view the terrain in three dimensions. Because the IMP camera on the lander had such a small field of view—only about fourteen degrees—looking at just one image at a time just wasn't going to be good enough. It would have been like wearing horse blinders and not being able to turn your head left or right. You'd always be wondering about what you couldn't see. So the CARD window was really six windows: six IMP images arranged three across by two down, all in stereo. The position of the rover in the terrain was represented by . . . the rover. Cooper had created a 3-D icon of Sojourner that could be turned or positioned anywhere in the scene; even its size would change, depending on how close or far away it had been placed.

Anyone who uses a personal computer is familiar with the mouse that allows him or her to move a pointer in two dimensions—left-or-right and up-or-down—to reach any location on the computer display. The RCW would control the rover in three dimensions, as well as specify the direction we wanted the rover to face. We needed a new input device. The Spaceball was the next step beyond the mouse. It looked like a baseball mounted on a post. The post rose out of a base that was bolted to the desk next to the keyboard. If you wrapped your hand around the Spaceball and pushed, the 3-D rover icon would fly directly away from you into the distance, getting smaller and smaller, until you couldn't even see it anymore. If you pulled, the rover came racing back, quickly filling the screen, then suddenly disappearing when it reached a position that was mathematically behind you. Push a button, and the rover would return to its "home" position, visible again. Push left or right on the Spaceball, and the rover slid left or right in the scene on the screen. The Spaceball itself didn't really move: It sensed the forces you were exerting on it. Try to twist it, and the rover turned, or rolled, or pitched forward. When you had the rover model positioned where you wanted it, you pressed another button to lock in a waypoint, the place you wanted the real rover to go. Cooper's sense of humor led him to design 3-D images of lawn darts embedded in the soil to show the user where those waypoints were in the window.

We continued our brainstorming sessions. The RCW evolved.

I'd say something like "I really need to know how long commands will take to complete. I need to know what time they're going to happen. Sometimes I'll need to know that time in Mars time. That way we'll be able to schedule an experiment at Mars noon if we want to. How can we do that?" Cooper would add to his list, and new features would in time magically appear in the software.

*

The first test of our ability to create rover command sequences on a tight schedule—as we would face with the rover on Mars—would come during early rover environmental testing.

Just as we subjected the Marie Curie to high accelerations and vibration to ensure that it would survive the launch and landing loads of the voyage to Mars, we also stressed it with "thermal/vacuum" tests to prove that the rover would handle the temperature extremes of the Martian surface without failing. The thermal and vacuum tests were combined because heat flows through a spacecraft differently in the vacuum of space than in the atmosphere of Earth.

The Marie Curie thermal/vacuum test was scheduled to begin on October 22, 1995. The plan was to take the rover through five Martian days of operation. Al Wen, the thermal engineer on the rover team, already had a mathematical thermal model of the rover, one that predicted how heat would flow as various devices were turned on and off and the temperature outside the rover rose and fell. But how good was the model? How fast would the modem heat up once its heater was activated? Much of the energy used by the devices onboard would end up as waste heat. (The WEB's design counted on it: This heat would warm the interior of the WEB enough during the day so that its contents wouldn't freeze during the night.) Would the electronics inside the Warm Electronics Box get too warm as the day progressed, forcing the rover to shut down early in the afternoon? The thermal/vacuum test would give Wen the data he needed to validate his model. And of course it would prove that the rover could survive the extremes of the Mars temperatures.

If there were any surprises during the test, there would still be a chance to make modifications in Sojourner to correct them.

✳

The plan for the test required that we exercise the rover in a realistic way. The performance of pieces of the rover might change with temperature. Unexpected interactions between those pieces might further impact the way the rover acted. And what we learned on one day of the test would tell the thermal guys what to change for the next day.

Henry Stone and Allen Sirota both had strong doubts about performing the test. They felt that the rover and the team just weren't yet ready. The test support team was pushing hard just to get the equipment ready in time for the reserved dates the test chamber was available. Stone in particular was concerned over how short a period Marie Curie had been available for performance testing before the thermal/vacuum test. Those of us on the software and operations team were only just beginning to learn how the rover worked as a system. There were plenty of software bugs left to uncover. Stone argued with Matijevic that if something went wrong with the rover while it was in the thermal chamber, we would not be able to determine whether the problem was a consequence of the thermal environment we had imposed on the rover, or due to a totally unrelated bug that happened to crop up during the test. Such bugs might mask actual thermal limitations. So if the team's experience was currently insufficient to distinguish these types of anomalies, what was the point in having the thermal/vacuum test at all? Perhaps it would be wise to postpone the test for a month or two.

Matijevic would have none of it. He did not disagree with the assessment that so far the team's overall experience operating Marie Curie was disquietingly small. But when he looked at the rest of the rover development schedule, he saw no future opportunities to repeat the thermal/vacuum test. The schedule was too tight. He was certain that there would be no second chance. The test would take place now.

Each day of the test, there would be an afternoon meeting in the room next to the chamber. The team would consider the progress of the test so far, and identify changes in the next day's plan. Layman, Eisen, Matijevic, the thermal engineer, and others would sit around the big table and discuss options. Eventually, we would go through the "strawman" command sequence, marking changes. I had insisted on a 3 P.M. deadline for those changes, since I would have to write the final commands, document them, and generate the coded versions using the Rover Control Workstation. But the test team often missed the deadline. I would sit in the meeting knowing that each minute of delay now would push me further into the early hours of the morning before I would be done producing the necessary command sequence for the next day. And of course I would need to come in to work early that day to ensure that the sequence was loaded properly and that the test could proceed. So as the planning meeting dragged on, my temper rose higher and higher, often evidenced by my snapping at the others at the table. When tomorrow's plan was finally agreed to, I ran out the door and across most of JPL to Building 107.

At the moment, the control station was still a prototype, barely ready to support these tests: Almost every command the rover could execute was available via the control station, but there was no mechanism for printing out the results in a human-readable form. As a result, I was forced to go back and forth between the RCW and the spreadsheet program I was using to generate the reports that the rest of the team could read. And since the RCW was still in development, it wasn't unusual for me to uncover bugs in its software as I was entering sequences. In some instances, the thermal/vacuum test was the first time the particular rover commands had ever been used in a sequence generated by the RCW. More than once the RCW software crashed when I tried to do something that neither Cooper nor I had tried out before, and I had to start over. If I encountered such a failing of the RCW late at night, I would call Cooper at home and we would try to reason out a work-around.

Marie Curie survived the thermal/vacuum test. We found no single problem traceable to the rover's fundamental design: We had proven that design sound.

Yet the testing had gone anything but smoothly. Sirota logged twenty-two new problems. There were problems with the early version of the rover's onboard software, causing certain commands to be unexpectedly skipped. There were problems with waking up the rover, making the rover stand up, taking APXS spectra, operating the WEB heaters, communicating, and even some minor hardware errors in the electronics boards that had not been caught previously. Many of the problems wouldn't have happened if we had had time to learn the proper steps to operating the rover. And, as a self-taught "student driver," I had put a few errors into my command sequences.

As Stone had predicted, nearly all of the problems would have been revealed in functional testing, if there had been time before the thermal/vacuum test. Fortunately, all of these problems could be corrected within the remaining rover integration and test schedule. Given the rover team's fears going into the test, I rated the results as Not Bad at All.

✳

Matt Golombek, the Project Scientist, had organized a field trip for some of the Pathfinder scientists and engineers. The trip would introduce the group to the Channeled Scabland of Washington State, a region on Earth that Golombek had determined was likely to echo the geologic history of Ares Vallis, the chosen Pathfinder landing site. Thirteen thousand years ago or so, a catastrophic flood had formed the Scabland, creating channels and carrying boulders and smaller rocks great distances, leaving them strewn around the region. The purpose of the trip was to give the team insight into geological formations the flight lander and rover might encounter during their mission. As a future rover operator, Brian Cooper participated.

During the Scabland trip the project also made one of its first attempts at outreach to increase the public's awareness of the mission, by inviting about a dozen educators to join the Pathfinder team in their travels. When the buses stopped at each site within the Scabland, the whole group would coalesce around Golombek while he described in detail the terrain features they were seeing, their geology, and how they had come to be. But other than including them in these discussions, most of the Pathfinder team wasn't quite sure what to do with the group of teachers that had been grafted onto the trip.

Cooper felt more of a kinship with the teacher group than with many of the scientists: "If I wasn't an engineer, I'd be a teacher. I love to explain things." And unlike many of the other engineers, Cooper wanted to talk to the group of educators, and find out their reactions to Pathfinder. "It was during that trip I met Fran O'Rourke," an elementary school teacher in Everett, Washington. When O'Rourke discovered that Cooper would be sending commands to the rover on Mars, she was filled with questions about how operations would be done.

By the conclusion of the field trip, O'Rourke had invited Cooper to speak to her elementary school class about Pathfinder and "driving the rover." Cooper immediately agreed: It sounded like fun, and he liked the idea of getting kids excited about science and space.

What Cooper didn't know was that O'Rourke had bigger plans for him. O'Rourke schemed with Cooper's wife, who then enlisted members of the rover team. We made sure Cooper had a rover to drive when he got to Everett.

"I brought Go-For 2, one of our prototype rovers." Go-For 2 was a simplified descendant of Brian Wilcox's fork-wheeled Go-For. "It was actually a purely teleoperated vehicle, had a camera and a video broadcast system. It was great for demos. It was great for showing kids what a rover could be like. So I brought that up."

Cooper had anticipated that he might end up talking to more than one classroom full of students. But when he walked into the auditorium of Cedar Wood Elementary School, the room was completely filled with people. Every class in the school was present. And many of the parents of those students were there as well, lining the walls. "They invited the press. It was a big deal for them. They had actually invited a lot of people from neighboring communities."

First Cooper spoke to the crowd, describing Pathfinder, the rover, and how he would operate the rover on Mars. "They had an incredibly smart group of kids there asking great questions. The students created posters, and maybe hundreds of things from all different grades, asking me what it would be like to drive on Mars," and telling him to be safe, and to drive with his seat belt on, as if he himself were going to Mars.

Then came the challenge. "They were going to award me the world's first Martian driver's license, if I passed the driving test. They had state representatives and the governor of Washington to officiate." A Washington State trooper oversaw Cooper's "driver's test." The trooper directed Cooper to put the rover through a series of maneuvers. "The kids had made these papier-mâché Mars terrains and had me drive Go-For around."

"Fortunately I passed," Cooper commented with mock relief.

The students had competed over the appearance of the driver's license, with many of them submitting designs. "It was a great way to get the school kids excited about our mission and space travel." The governor presented Cooper with his Mars driver's license, far too large to carry in a wallet, but suitable for framing. Another copy of the Mars driver's license was sent to the Smithsonian collection in Washington, D.C.

In time we were using the Rover Control Workstation to command the Rocky 4 rover in the Building 107 sandbox. The leftover MSM simulated lander, sitting in one corner, took the place of the IMP camera. Cooper donned his goggles and designated lawn dart waypoints; then we watched to see how well the rover did actually getting to the target.

Donna Shirley periodically nagged the rover team about field testing the rover. She didn't trust the results we got running Rocky 4, and later Marie Curie, in the sandbox. Even driving through the new outdoor test course—dubbed the "MarsYard" and painstakingly created by the rover technology program—was not enough. Shirley would be placated only by a true field test, with Marie Curie navigating through natural terrain somewhere out in the desert.

She also pointed out that Hank Moore, the geologist who had been chosen as the Rover Scientist for the Pathfinder mission, thought stereo images could be misleading. He'd been part of the Viking team many years before, and had studied stereo pictures sent back from Mars by the landers of that mission. Moore wasn't sure you could do a good job estimating where the rover was: It seemed to him that the rocks in the Viking images had all been closer than they appeared. Shirley echoed Moore's concerns, and wondered when we would finally take the rover out to a "real" location.

When I first heard of Moore's warning, I bristled. I figured Hank Moore might have credentials as a scientist, but not as a rover operator. Estimating locations precisely in stereo images was exactly what we had implemented CARD to do. We had already demonstrated that it worked several times. Who was this guy? At the time, "Hank Moore" was just a name to me; I didn't know that getting to know the man would be one of the great privileges of being on the Pathfinder mission.

Field testing was going to be a logistical headache. And we were still uncovering subtle problems even with our sandbox tests. Why go out to the desert before we were ready? Shirley prodded. I protested. This interchange seemed to occur every few months. It usually ended with me promising that the field test would happen when the time was right, and Shirley acquiescing, saying something like "Just so long as you get around to it."

＊

In September 1996, three months before launch, Brian Cooper and two Rover Control Workstations moved from Building 107 to the Space Flight Operations Center. By now, the RCW had become a solid system. Almost all of the bugs had been slain. There would still be tweaks and minor additions to the software, because experience would lead the rover sequence builders to ask for new features. But the RCW was already the workhorse of our systems and operations testing, just as it would be when the "tests" were on Mars.

The second floor of Space Flight Operations Center was in flux. Most of the Pathfinder operations team was already there. Some of the engineers who had worked on the development of the lander's hardware and software were moving out, making room for those who would fly it. Office cubicles were going away, changing shape, re-forming into huge conference rooms and an open "bullpen" where the science teams would set up shop. Pathfinder's Mission Control room, called the MSA, for Mission Support Area, was already in place in the center of the second floor.

Blue signs with white letters went up in the MSA and elsewhere around the second floor to identify key mission functions. The sign over Cooper's door read simply "ROVER CONTROL." The office was big—it would eventually hold three or four members of the rover uplink team, working furiously together to plan rover activities and transform them into the right commands. All the commands Sojourner would ever execute would originate in this room.

More construction paper greetings and messages arrived from the students of Cedar Wood Elementary. Cooper tacked them up on the bulletin board in the hallway outside his office, so that everyone working Pathfinder could see the excitement their mission had already engendered.

Jake Matijevic, now the Rover Manager, had his office down the hall. Within a couple more months Henry Stone and I had moved our offices to just around the corner from the rover control room.

Brian Cooper surveyed all the activity. "We're really going to do this!"

EIGHTEEN

METEORITES, LIFE, AND JOB SECURITY

On August 7, 1996, scientists at a news conference in Washington, D.C., announced that possible evidence of life on Mars had been discovered in an Antarctic meteorite. Analysis of the rock had determined that it was indeed of Martian origin. The science team studying the meteorite, which had been designated ALH84001, was very careful to state that they were not claiming to have found proof of either existing or past life on Mars. They only said that there were a number of separate pieces of evidence that could together be plausibly explained by a biological process, and only somewhat less plausibly understood through a combination of chemical reactions. There were even tiny shapes, visible only in microscopic images, that the scientists proposed "might" be micro-fossils.

The media wanted to know what the impact of the news would be on the Pathfinder mission. The answer: none. The press found this perplexing. The possibility of life on Mars was a big story. How could the next mission to the Red Planet ignore what had just been discovered?

To the engineers working on the Pathfinder project, the excitement over meteorite ALH84001 reflected an intriguing discovery about the planet they were targeting. And if life truly had been discovered on Mars, it would spell "job security" for every NASA employee.

But Pathfinder had never been intended to search for life. The hardware of the lander and rover was complete, and was only a week or so from being shipped across the continent to Kennedy Space Center for integration with the launch vehicle. The designs had been fixed for almost two years and were now immutable.

The response to the possibility of life on Mars would have to wait until future missions.

Yet JPL and NASA did not ignore the firestorm of attention and interest now focused on Mars. The Mars Exploration Program, led by Donna Shirley, had already been studying the possibility of a Mars sample return mission that might launch as early as 2005. Now, with the President of the United States calling for an international scientific conference to determine the appropriate next steps in response to the evidence of past life on Mars, NASA headquarters wanted to know if the schedule for Mars sample return could be pushed forward. Could a mission be ready to launch by 2001?

A small group of engineers at JPL found itself running around in a mad rush to try to answer that question. Most of the rover team was blissfully ignorant of this activity. But Jake Matijevic was the MFEX Rover Manager, and could not stay out of the fray. He was the only one available who could imagine what a rover that could collect Martian rocks would look like, and what it would cost to create. Matijevic shielded the Sojourner rover team so that they could make today's mission—one built of hardware instead of paper—a success.

*

For the final two weeks before Sojourner left for the Cape, the rover schedule showed fifty-nine separate tasks to be completed. Most of these were tests, verifications, and calibrations of the flight rover. The rover team worked to a plan mapped to the day, sometimes to the hour. Weekends were left open, and marked as "contingency days." If we got behind schedule, we would be working seven days a week to catch up.

The effect of the pressure was pervasive throughout the rover team. We all tried to make allowances for each other's peccadilloes and lapses. We were still held together by our shared objective to put a working microrover on Mars. Some people handled the stress better than others.

Working long hours was like an addiction. Once I'd done it for long enough, things just didn't seem right unless I was working. As we made the last push toward shipping the flight rover, I often found myself angry and snapping at people on the rover team for no good reason.

It was time for a vacation.

✳

Once Sojourner was on her way to Kennedy Space Center, I took a full week off. For most of that week I did next to nothing: I slept late, read novels, watched TV.

By Friday night, I was relaxed enough that I felt like tinkering on my home computer with a rover command simulator I'd been working on in off-hours. The simulator was my pet project, a spreadsheet program that would let me predict how long the rover would take to complete the commands we gave it, and how much data it would send back as a consequence.

On this particular Friday night, I thought I'd tackle a command I had not yet implemented: the "Local-Time WAIT." We had several ways to tell the rover how long to sit idle. There was the "Relative-Time WAIT," which we used whenever we wanted the rover to wait a period of time between completing one action and starting the next: "Turn on this sensor, then wait five minutes for it to warm up." We could also tell the rover to wait until a specific day, hour, minute, and second.

The "Local-Time WAIT" command let us tell Sojourner what to do according to a Mars time clock. Since the Martian day was thirty-nine minutes longer than an Earth day, the Earth clock and Mars clock were always getting out of synch with each other. It wasn't like the constant difference between two time zones on Earth; instead, the time zone difference kept changing, day by day. Specifying times according to an Earth clock would have been less convenient for scheduling rover activities that were naturally tied to the Martian day, such as observations that would always be done at noon, when the sun was as high overhead as possible.

The rover also had a built-in safety feature—auto-shutdown—that depended on the Mars local time. If solar power dropped below a set level, and the time of day was later than specified (usually 6 P.M. Mars local time), the rover would shut itself down, even in the middle of a command sequence. The idea was to prevent the rover from inadvertently running down its precious, nonrechargeable batteries by operating for long periods at night.

I wrote a short software routine to do the "Local-Time WAIT." I tested it with a "Wait until 10 A.M." command, but it wasn't working. Thinking through Mars clocks and Earth times was a bit tricky, so I hadn't expected the routine to operate perfectly the first time. I went through the software code again . . . I wasn't finding the problem. The software routine just didn't have that many lines of code in which a bug could hide. What I had planned as a short exercise was taking a long time. The more I studied the software, the more I was sure that I had implemented it correctly. So what was wrong? I was starting to get a bad feeling. The one piece of information my software depended on to be able to tell time— Mars local time—was the Greenwich Mean Time that matched midnight in Mars local time. By telling the rover when midnight was on Mars, it could then compute when any other Mars local time would occur. If my software was correct and the results were wrong, that could only mean one thing: The stored time specifying Mars midnight was wrong. My simulator was using the same time values that I had loaded into Sojourner's memory, so if the time was wrong in the simulator, it was wrong on the flight rover as well. I checked the value. Sure enough, the time was incorrect. I changed the number to true Mars midnight, and the results in my spreadsheet immediately made sense.

I thought, "What a screwup!"

As I studied the error, a horrible scenario came to mind. The time for midnight that was currently stored on Sojourner was so far off that early morning on Mars would seem to the rover to be the middle of the night. I could imagine the rover waking up at around 8 A.M. Mars local time on landing day, noting that the solar power level was low and that the local time was after its scheduled bedtime, and then shutting itself off—all before the first command sequence came from the lander. Sojourner wouldn't know that the sun had just come up, but would instead think the sun was setting. If the operations team couldn't get enough telemetry from the rover to determine what was going on in the short time that the rover was awake, it might take days to understand Sojourner's sudden proclivity to power itself off, an ironic reversal of the spurious wakeups we had experienced during early testing of Marie Curie.

Of course, I'd found the error. The scenario would never happen. But I couldn't stop wondering what would have happened if I hadn't decided to play with the simulator. When would I have caught the problem? Would anyone else have ever come across it?

When I went back to work the next Monday, I rushed to announce my mistake to the team. I was then curiously disappointed when my midnight error didn't even rate an entry in Allen Sirota's problem log. He and the rest just didn't see the midnight error as a problem: It was simply a parameter change to be made. From Sirota's point of view, the rover had never exhibited the symptom, so there was nothing to report. To me, it was a time bomb waiting to go off.

A month later, one extra command was added to one of the prelaunch sequences sent to Sojourner at KSC. The parameter in the rover's memory was changed to the correct value for midnight on Mars, and that bomb was defused forever.

✳

For the first leg of the trip to Mars, the rover and lander traveled separately. The lander, intended to fly through space, was trucked from Pasadena to Cape Canaveral, Florida, arriving in mid-August.

Two weeks later the rover, designed to drive, was flown across the country by commercial airliner. Sojourner had its own custom storage container, which the rover team commonly referred to as the "sarcophagus." The words "MARS ROVER" were printed on each side. In preparation for the trip, a standard airline modular cargo container had been delivered to JPL. Eisen and his crew loaded the sarcophagus into the cargo container and carefully strapped it in place. They transported the cargo container to the airport, then went out on the tarmac and supervised the loading of the Mars-bound cargo into the aircraft. From there, airport regulations forced Eisen and Ken Jewett to return to the terminal. They walked through Security and boarded Sojourner's plane with the rest of the passengers. After the flight landed in Orlando, Eisen and Jewett witnessed the unloading. Sojourner was three thousand miles closer to Mars.

✳

When Tom Economou arrived at Kennedy Space Center with the flight APXS sensor head, it didn't work. The flight head was supposed to be identical to the spare unit already installed on Sojourner, except that it contained the radioactive sources that the APXS needed to get meaningful data. Allen Sirota was getting used to dealing with Economou. "Sure enough, it didn't work. I mean, it didn't work at all. Nothing but noise in the spectrum." So they transferred the radioactive sources from the flight sensor head to the spare unit. The spare was remounted onto Sojourner, and the APXS started working again. "The identical sensor head— wasn't." The spare unit would be flying to Mars.

For the last time, the mechanical team tied Sojourner down to the lander petal. The RHUs were installed. Ken Jewett carefully locked down each wheel. He took off all the components with red tags saying "REMOVE BEFORE FLIGHT," ensuring that no lens caps or laser covers would inadvertently make the trip to Mars. Jewett was glad to be done with Sojourner. For him, one of the most satisfying moments of the project had been back at JPL, "putting the rover together for the last time . . . knowing that it was the last time." Over the past year they had had to tear down Marie Curie and Sojourner, and then reassemble them many more times than anyone had expected. The final assembly occurred in the early hours of the morning. There was no one else around. When it was done, Jewett stopped to appreciate the moment. "I just let the building quiet down. Every cable inside and outside was tied up exactly where we wanted it. Nothing on the vehicle was out of place. There would be no sleepless nights wondering if we'd made any mistakes. It really went together well."

At the Cape, Jewett watched the Pathfinder team close up the lander. The next time anyone would see Sojourner, it would be in images from another planet. The lander was mated to its heatshield, backshell, and cruise stage. Later, Jewett found himself "watching it go on the rocket.

Knowing that there was nothing more I could do." Jewett paused. He smiled. "And I would never have to take it apart again!"

※

The prelaunch party was held at the Mission Manager's house. People toasted to a job well done. They talked about where they would be during the launch—at JPL, in the blockhouse at Kennedy Space Center monitoring the spacecraft, or out in the bleachers just watching. They seemed excited, relaxed, happy. This would be the last chance for most of the team to gather together in one place, before they scattered either for a well-earned Thanksgiving holiday or to participate in final launch preparations on both coasts.

I stood in the corner, next to the food, and spoke with Miguel San Martin and his wife. San Martin was a tall, blond, balding, expressive Argentinean. He worked the ACS—Attitude Control System—for Pathfinder, the software that made sure the spacecraft was pointing in the right direction, which was critically important when firing thrusters. San Martin was comparing Pathfinder to other projects. "It's amazing, really. I was working on flight software for Cassini. We were going to Saturn. And they managed to make it boring! Everything was compartmentalized. You could only work on your one area. There was no opportunity to apply your skills anywhere else. If you discovered a problem, no one was interested in letting you deal with it yourself." On Pathfinder, you just did what had to be done.

Although mostly working separately, and often at odds over specific issues, the rover and lander teams had shared a dedication to success. I believed in both teams. I was confident in the ability of every team member I had worked with. But when you pause for breath after running for so long, questions you cannot yet answer have time to form. Would it work? Had any one engineer missed something vital?

Had we really done everything that needed doing?

PART 3

GOING TO MARS

NINETEEN

EVEN A JOURNEY OF
A THOUSAND MILES . . .

The day before Thanksgiving, four days before the appointed hour, we already knew we had an 80 percent chance of scrubbing the first launch attempt. The rover was ready. The spacecraft was ready. So was the launch vehicle. But the long-range weather forecast showed a front moving in at Cape Canaveral. Rain and high winds would erode our safety margin. The project could not risk the rocket being destabilized like a motor home being battered around on the highway in gusty winds.

One unchangeable rule of solar system exploration is that the planets don't wait. They move inexorably in their orbits and cannot be cajoled into adjusting their schedules to match those of mere humans. Only through cleverness can we leverage the limited lift capacity of our rockets and the occasional alignments of the planets into opportunities to send spacecraft across interplanetary space to actually find a planet at the other end.

However, with some extra propellant in the tanks, you can buy some leeway. We had a chance to launch once each day from December 2 to December 31. After that, we would have to wait another twenty-six months until Mars and Earth were in position to try again. Realistically, NASA was not going to pay to keep the entire operations team active for two years doing nothing. So if we did not launch in December, we might not launch at all.

For interplanetary missions, launch windows—the periods of time each day that a launch will get the spacecraft where it needs to go—are critical and fleeting. During the first two weeks of December, there was a two-minute launch window each day. After that, the window would rapidly shrink down to only one second each day. By January, the energy capacity of the rocket would just not be sufficient to get to Mars.

When Pathfinder was launched, it would first go into orbit around the Earth, and then be given a final boost to put it on its way to Mars. If only the Earth's equator were exactly aligned with the plane of the solar system, then Pathfinder could be launched at any time of day, and boosted into its Mars trajectory whenever it reached the point in its orbit when it was headed in the right direction. But the Earth's axis is tilted about twenty-three degrees from vertical. So to make full use of the speed advantage of the thousand-mile-per-hour spin of the Earth, we needed to launch only when the Earth's motion would hurl Pathfinder along the plane in which all the planets traveled. And this happened only for an instant, twice per day.

For half the day, the Earth's spin would hurl Pathfinder out above the plane. For the next half of the day, it would hurl Pathfinder out below. Only during the brief period of crossover would the spacecraft be hurled perfectly along the plane, on target for Mars. Those short moments were the launch windows.

There was yet more complexity in the planning of a launch. When a space probe is sent to Mars, it is not a simple case of "point and shoot." The current position of Mars is not the target, for by the time the spacecraft gets there, Mars will surely be gone. Instead, the spacecraft must set out on a path that will cross the orbit of Mars at a specific point and at the precise time that Mars will arrive there. If you can master that objective, the next problem is slowing the probe upon arrival so that it doesn't make a crater when planet and spacecraft meet.

✴

The first launch window was a two-minute period centered on 2:09:11 A.M. EST, December 2. By the morning of Sunday, December 1, the probability of being forced to scrub due to weather was up to 90 percent, so the decision was made to hold off until the next launch window, on December 3. The weather forecast said that, by the third, conditions would be much improved, with an 80 percent chance of a "go" for launch.

Since Sojourner would sleep through the launch, there was nothing for our team to do during the final preparations. Only a few members of the rover system team had to be in Florida in case a serious problem with the lander/rover system was identified before launch. Nevertheless we were all in town. For most of us, this was to be our first launch. All of us wanted to see Sojourner on its way.

With the launch postponed one day, the postlaunch reception on the patio of the Patrick Air Force Base officers' club became another prelaunch party. The base was only a few miles south of Cocoa Beach, so you could look north over the water to where the Delta rocket shone bright white in the crossed floodlights that illuminated it. Beacons on either side of pad C17A blinked red, signaling a payload ready to be launched. We were miles away, and yet the reason for the party was there to be seen, poised to begin its journey.

The waves rolled onto the beach, barely visible in the light spilling over from the patio. I stood there and talked with Jack Morrison, looking for Mars. After a bit we thought we had found it, rising above a low cloud bank. Just a star with a red tinge to it, a point of light a hundred million miles away. Tonight we were going to send something there!

I looked from Mars to the rocket. Back to Mars. Back to the rocket. The origin of the trip was in view. So was the destination. Music played behind me. The starting gun was set to be fired in just seven hours.

✴

The launch was scheduled for 2:03 A.M.

There were three viewing sites: the VIP bleachers, the NASA Causeway, and Jetty Park. The bleachers and causeway were within the confines of the Cape Canaveral Air Station, north and west of the pad. You needed a car pass to get onto the base; they wouldn't even let you in until one hour before launch, and people would be jockeying for position in the spots they were allowed.

Since the launchpad was near the southeast corner of the base, Jetty

Park—outside the base, where anybody could go—was actually the closest site to the pad. The morning before the scheduled launch, we had scouted the area. The jetty had been built of rocks piled on each other. It reached a few hundred yards east out into the water, topped with a narrow concrete walkway railed on both sides. Periodically the walkway widened to make room for sinks—so that the fishermen who lined the jetty day and night could clean their catch—then narrowed again. In daylight, we could see pelicans sitting watch from every available piling. When they took flight, it was to gather around a pair of porpoises making their way out to sea. A colony of feral cats seemed to be living among the rocks of the jetty itself. As we approached, they grudgingly gave ground, seeing us as interlopers on their turf. The fishermen ignored us.

From the jetty, only the bottom third of the rocket was obscured. And because the jetty was so long and narrow, it was unlikely that anyone would block our view.

The rover team had agreed that Jetty Park was the place to be. At midnight, the convoy of rover team members, family, and friends headed for the jetty. The line to get in was short, but there was a line. They were charging the usual dollar per car to park. I wondered distractedly whether there was someone there at midnight on non-launch nights to take your money.

There might have been a hundred people out on the jetty before us, but there was plenty of space. We established our personal spots among the others. Jack Morrison and his wife had brought a pair of binoculars. We passed them around, taking turns studying the Delta rocket. It was beautiful. To the naked eye, the Delta had just seemed a tower of white. Through the binoculars, you could see that it was actually mostly a grayish blue. On it were the symbols of Pathfinder, and of McDonnell Douglas, manufacturer of the Delta.

We waited. It was cold and damp. Yet some people were in T-shirts. Nervous energy, I supposed.

We waited. We talked. I paced. We told stories, thought about the Russians' Mars '96 mission that had ended at the bottom of the ocean only two weeks before. Mostly, we were nervous. With excruciating slowness the two-minute launch window was grinding toward us.

A groan spread through the group. "They scrubbed the launch! A console! A damn console!" The crowd began to disperse. I kept staring at the rocket. I didn't want to believe it wasn't going to happen. But after another moment, I knew the launch window must have closed. A minute before the launch, a software glitch had occurred in a ground computer monitoring telemetry from the Delta. Not enough time to fix it before the window expired.

Tonight would not be the night.

✳

We were back the next night. But there were fewer of us. The fishermen were grumpy about all the people scaring the fish. For tonight, the launch window would be slightly earlier, at 1:58 A.M.

The Principal Investigator for the APXS came by. "So, Andy. What's the reason going to be tonight? Why won't we launch this time?" I told him there would be no reasons, that tonight we would launch. I hoped I was right.

About a minute before the launch time, people got quiet. I heard someone reciting a countdown. We all stared at the Delta. Then the base of the rocket got suddenly brighter. And the rocket was moving. "That's it!" a voice shouted. Everyone cheered and applauded. Almost everyone. I could only watch. I willed the rocket into the sky.

There was no sound yet from the rising rocket. It was too far away. It climbed into the sky, so bright that I thought I should look away, but didn't. All the clichés about "bright as the sun" were true. I could see a region around the flame in my vision where my retinas were saturated. A small part of my mind wondered if I'd see all right after this was over.

The crowd spontaneously cheered again. I wasn't sure why, except maybe that it was clear that the rocket was flying true. The sound reached us finally across the water. A kind of staccato roar, voice of pure power. I couldn't see the Delta itself anymore, just the flame and the trail of exhaust that drew an arc between it and the pad. It looked like it was aimed straight at the moon.

The flame was smaller now, but it seemed just as bright. Then it stuttered briefly, dimming and brightening again. There was the shortest pause, and the six ground-lit solid rocket boosters became visible as they peeled away from the Delta, their work complete. The three air-lit boosters had taken over. Another cheer.

The flame of the engines was rapidly becoming a bright red star in the sky.

The jettisoned solid rocket boosters had been left behind. They formed a flickering constellation. I knew that they must be on their way to falling into the sea, but for the moment they seemed to be hanging in the sky, six twinkling lights, a new Pleiades.

Only one minute later, the air-lit solid rocket motors were jettisoned as well. But the Delta was already too distant for me to discern the boosters falling away, though some others in the crowd claimed they could. Just two minutes into its mission, the spacecraft was traveling over a mile and a half per second.

The spot that was Pathfinder was tiny now, hundreds of miles away. I lost sight of it.

The champagne bottles were coming out. Some people had tears streaming down their faces and didn't even realize it. The rover crew gathered as most of the other watchers headed back toward their cars. We all shook hands. "Good work! Good job!" everyone said. Could I really take credit for what I had just seen? I wasn't sure.

Allen Sirota raised a cup. "To the rover!" We all drank the toast.

We started packing up to follow the rest back along the jetty. Again we all shook hands. I found myself repeating the same statement to everyone: "We're in business!"

With Pathfinder on its way to Mars, many people's jobs were complete. Mine was just starting.

TWENTY

CRUISIN'

Pathfinder was on course for a Fourth of July landing on Mars. The cruise stage was spinning at twelve rotations per minute, as planned. But by the time we got back from Florida, the spacecraft was already in trouble: When first activated, the onboard sun sensor, critical to navigating the spacecraft, was half-blind.

The sun sensor was there to measure the rotation rate of the spacecraft, and its approximate orientation in space. The sun sensor had five sensor heads situated on the body of the spacecraft, facing in different directions. Sensor heads 1, 2, and 3 faced to the sides of the spinning cruise stage; heads 4 and 5 looked back along the spin axis. There was also a star tracker onboard, which would determine the rotation rate and orientation more precisely, but only once the sun sensor had provided the computer with its rough estimate of this information.

To make course corrections during the cruise to Mars, the spacecraft computer would require precise knowledge of the direction the spacecraft was pointed. Firing thrusters in the wrong direction would just make things worse, and would cause Pathfinder to miss Mars completely.

The software team manager explained the impact of the sun sensor situation matter-of-factly: "We're not running the Attitude Control System now, and we can't until we can use the sun sensor. We're almost certain that there's debris obscuring the optics of sensor heads four and five. If we could send somebody out there with a tissue and some alcohol, we could fix it in a minute. Not an option. If we can't fix the problem, it's the end of the mission. Do I think we can fix it? Yes. We're working the problem. We think we have a solution. If it works, everything is okay. Otherwise, we can all go home."

In the meantime, since the Attitude Control System, or ACS, could not be turned on without the sun sensor to guide it, team members in other subsystems identified alternate means to determine the spacecraft spin rate and orientation. By examining the change in the strength of the radio signal transmitted from Pathfinder as it rotated, the telecommunications engineers came up with estimates for both. The planetary navigation team derived estimates of the spacecraft's distance and speed by sending a radio signal to the spacecraft and waiting for the precisely timed response.

The telecommunications and navigation representatives presented their results each morning at the status meeting. Though independently generated from separate data sources, the estimates of both teams were a nearly perfect match.

The story began circulating through the ground team that everyone knew more about the orientation of the spacecraft than Miguel San Martin, who was the engineer responsible for the ACS. "Even the janitor knows more about the spacecraft attitude than Miguel" went the joke. Of course, San Martin didn't yet have a working sun sensor to rely on. It may have rankled some that two other subsystems were reporting the information that should normally be coming from him, but San Martin took the situation in good humor. In fact, he was the first person to tell the joke. The data from the telecommunications and navigation teams did allow the project to track the state of the spacecraft, but it was useless for attitude control, since the estimates were after-the-fact determinations based on ground processing of large data sets. Eventually the spacecraft would be so far away that any instructions sent from the ground would take over ten minutes to arrive onboard. For course corrections and maneuvering, the spacecraft needed to assess its own orientation and spin onboard in real time. And that required the sun sensor.

Since the internal electronics of the sun sensor was functioning normally, the theory was that debris was blocking the sensor heads. The most likely source of the debris was contamination from the protective shroud that covered the spacecraft during launch. Once in space, the two parts of the shroud had explosively separated, as intended, falling away to either side of the Delta. Perhaps during separation some propellant had spattered onto the sun sensor heads.

Attitude control for Pathfinder could not be shifted to depend on only the remaining functional sensor heads. Even though there were five heads in all, they were not truly backups to each other. Each head was pointed in a direction different from the others. Working in concert, the sun sensors would track the sun with the spacecraft in any orientation. For most of the trip to Mars, the spacecraft had to be oriented with its solar panel in the general direction of the sun. Only sensor head 4 or 5 could do the job.

Sensor head 5 was gone. Or at least that had to be the assumption, since it was generating no signal. The chance of it spontaneously cleaning itself seemed remote. Similarly, head 4 was not likely to improve. So the Attitude Control System would have to function with what head 4 could now provide. Careful review of the data from head 4 confirmed that it was putting out a signal that looked normal, except that is was four times weaker than it was supposed to be. Even such a weak signal should have been enough to track the sun. But the onboard software was refusing to use the head 4 data. This was a designed-in safety measure: If the magnitude of a sensor head's output dropped too low, it was considered suspect, and the software rejected the data in favor of that available from the other heads. This made sense in the case of a problem with a single sensor head, but with head 5 out of the running, there was no other data to fall back on, and the safety protection feature prevented the Attitude Control System from operating on the only good data it had.

The lander software team wrote a modification to the onboard software, a "patch" that would cause the Attitude Control System to accept the weak signal from sun sensor head 4. This patch would replace a small piece of the program running on the spacecraft, to change the conditions under which sensor data would be rejected. The new software was first tried out in the spacecraft testbed. With the weak signal simulated, and all other aspects of the spacecraft configuration duplicated with all the fidelity the ground team could muster, the patch functioned as expected in the testbed.

On December 7, four days after launch, the patch was to be uploaded and installed on the actual spacecraft. It was a Saturday, and a small group came into JPL to handle the job. No one expected any problem: They'd just uplink the data files containing the patch, install the patch, and start running the Attitude Control System. The whole process should take two hours.

But when the Pathfinder team uplinked the files, they didn't go through. After repeated attempts, the spacecraft continued to reject the files. What was going on? They determined that very short data files and individual commands were being received onboard, but nothing else. More and more members of the Pathfinder team were showing up as word of the problem got out.

They tried slicing the original data files into many smaller files, and sending those. This was partially successful: Some of the files arrived onboard, but others did not. They still didn't understand the cause of the problem. And it was getting perilously worse: Sometimes even tiny individual commands weren't reaching the spacecraft! If they couldn't command the spacecraft, the mission was over.

The answer finally came from the telecommunications engineers. They had been monitoring an unexplained dip in the strength of the communication signal from the spacecraft, a short dip that recurred every five minutes. Now what would cause that? The spacecraft was rotating at twelve revolutions per minute, or once every five seconds. The measurements they had been taking of signal strength weren't continuous; the telecommunications engineers realized that the dip was actually happening every five seconds, but they were only catching it in their data every five minutes. Some piece of metal was blocking the radio signal once every time the spacecraft rotated. No wonder the files weren't getting through: They were getting chopped off in the middle of transmission.

To get data up to the spacecraft, they applied the usual technique for dealing with very weak radio signals, one that should never have been needed with Pathfinder still so close to the Earth: They dropped the data rate. The data files were uplinked at 7.8 bits per second, one-thirtieth the normal rate for this point in the mission. This time they got through. The patches were installed.

After a harrowing twenty-hour delay, the command was sent to activate the Attitude Control System, and it was up and running for the first time in the mission. The Pathfinder team commanded the spacecraft to turn a bit more toward the Earth, and the communications problem went away.

A few days later the spacecraft was "spun-down" from 12 rpm to 2 rpm. At this rotation rate the star tracker would operate. With the spin rate computed from the working sun sensor output, the ACS was shifted into "Celestial" mode. The star tracker began scanning, and locked onto two stars. Pathfinder now knew exactly where it was pointing.

Miguel San Martin was once again the principal source of spacecraft orientation status reports.

❉

Sojourner had ridden through the sun sensor campaign in oblivious sleep. The first in-flight healthcheck, through which we would verify that the rover had survived the launch unscathed, was planned for December 17, about two weeks after launch.

This was going to be our first moment of truth.

There was no reason to expect that there would be anything wrong. Ever since the sun sensor fix, I had been relaxed. I had convinced myself that the only times to be worried were launch and landing. It seemed to be working. Until Allen Sirota came by a few days before the healthcheck and talked about being nervous. Other engineers dropped by with their own ideas about which rover components might fail. So about twenty-four hours before the rover wakeup, my stomach started churning. What if we ordered the lander to power on the rover, and all we heard was silence? What would we do then? How could we fix it without any data?

Every member of the rover team had shown up for this first postlaunch check of Sojourner. They each knew how their particular subsystems functioned, how they were built. But not all of them knew yet how to interpret the full telemetry stream. I warned them: "If everything goes as it's supposed to, we should receive exactly one 'Error Report' message. This is not a problem. It's not really an error. When the rover wakes up, it will check the accels and note that it's not seeing any gravity. That will trigger an automatic mission phase change from 'prelaunch' to 'cruise' phase. Every time the rover does a mission phase change on its own, it sends a telemetry report to tell us what it did, and that report shows up as an error message." Someone asked which type of error we would see. All error types were numbered in hexadecimal code. I looked it up. "1D02." If we saw a different error message, or a second one, then we would have problems.

For the telemetry that would be generated by the healthcheck command Allen Sirota and Art Thompson had polled the various subsystems and created a checklist of the numerical values expected for each of the hundreds of data channels, and how far the channel could deviate from the expected value without raising an alarm. The expected numbers were based on the values observed prior to launch, together with the few changes expected due to the postlaunch environment. The primary changes we anticipated were in Sojourner's onboard temperatures, and in the rover's gravity readings, which should be virtually zero, since the rover was in free fall on its way to Mars.

Just powering on the rover was more important than the healthcheck itself. The change to "cruise" phase was the most critical step between now and the landing on Mars. All we needed to do for the rover to know that it was no longer on the Earth was to wake it up sometime during cruise. Just to be doubly safe, we also told the rover to switch to "cruise" in the command sequence for the healthcheck.

Once the rover was in "cruise," it would be ready to switch over to "on-Mars" whenever it next woke up and sensed gravity again. So even if the radio modem was damaged in the landing, Sojourner would still do the right thing and perform its preplanned contingency mission. However, if the rover was never powered on during cruise, and the modem failed, Sojourner would wake up on Mars still in its "prelaunch" phase, with no way to determine that it was not still on Earth. In the wrong mission phase and with no communication, it would sit on the lander for all eternity, or at least until some future Mars astronaut picked it up to take it back to the Smithsonian.

Art Thompson was the rover engineer assigned as liaison with the project team for the day's healthcheck. I sat with him in the Mission Support Area, the Pathfinder equivalent of Mission Control. The room contained a dozen Sun computer workstations, and VOCA (Voice Operated Communications Assembly) units that enabled the operators at each station to communicate with each other over the voice network. The rest of the rover team was situated down the hall at their own Sun workstations with appropriate telemetry displays. They were also hooked into the voicenet, but it was Thompson's job to represent the rover to the Pathfinder ground operations team. The operations team would refer to Thompson on the net simply as "Rover."

As the team worked through each step of the procedure to prepare the spacecraft to support the rover healthcheck, the Flight Controller polled each position to ensure that we were unanimously ready to continue.

Back in November, we had uncovered a bug in the lander's software that caused the computer to reset—temporarily bringing down the spacecraft—whenever the Attitude Control System was running and the lander and rover were communicating. Now we reached the step in the procedure where the ACS would be shut down to avoid this problem. Miguel San Martin was at the Attitude Control System position. "ACS is go."

The flight engineer also confirmed he was ready to send the instructions. The Flight Controller approved the uplink, and off went the commands, through the Deep Space Network and across 2 million miles of space. Telemetry came back, first confirming that the spacecraft had received the command, then indicating the command had been executed. San Martin reported through the net, "ACS confirms we are now in super-idle mode. ACS is offline. No problems."

Richard Cook, the Mission Manager, had been standing in back watching the goings-on. "Okay, Miguel. Everything is fine, and ACS is shut down." Off the net, he continued mischievously, "Without ACS we don't need you. You're fired. Go home."

San Martin stayed. If we followed the procedure, ACS would be running again in fifteen minutes.

Slowly, carefully, we worked our way through the procedure until we were ready to run the lander command sequence that would wake the rover and send it commands. "Rover is go for sequence activation."

The command went up. "Don't reset us, rover!" commented the Flight Controller, to the ire of the rover team. It went across the voicenet, probably unintentionally.

The lander sequence turned on the lander-mounted rover modem, the lander's side of the communications link to the rover. We knew the sequence would wait ninety seconds and then cause the lander to apply power for ten seconds to a magnet located just under the reed-relay on the rover. The relay would activate, and the rover would begin to power-up using its own onboard battery. The rover would take about thirty seconds to go through its startup process, then send its initial data. At the speed of light, a message would require about thirteen seconds to cover the distance to the Earth . . .

There was the error message. Sojourner lives! "We've got rover telemetry!" It was the right error message: 1D02. "Rover is now in mission phase 1, cruise phase. There's the healthcheck. We're preparing for APXS operation. Rover's now in a comm hold for the APXS. The next command we see should be number 2010, then 2009. There it is. Out of comm hold. Now we're reading out memory. Yes. Yes. Rover has shut down!"

We had just had a conversation with our rover, which was almost ten times farther away than the Moon, and it had worked just like a ground test.

While Miguel San Martin oversaw the reactivation of the spacecraft ACS, the rover team met to review the telemetry in detail. All the voltage and current readings matched. All eighty devices onboard the rover reported status as "Good." Some temperatures were lower than we'd thought we would see. But our expectations turned out to be based on the lander's thermal environment from yesterday, not today. So our expectations were wrong, and the rover's telemetry was right. Just the kind of discrepancy we wanted to have. All the voltage and current readings matched.

With the healthcheck complete, I could write the first rover status report for the Mars Pathfinder telephone information line:

This morning the operations team woke up the Sojourner rover for the first time since launch. The rover performed an internal health status check, accepted command sequences provided by the lander, operated the onboard Alpha Proton X-ray Spectrometer, sent telemetry, and shut itself down as planned to conserve its batteries. We are happy to report that the rover is healthy, with all subsystems functioning normally. The rover will remain powered off until the next rover healthcheck, not long before landing on Mars.

When I got back to my office the next day, I found a message on the whiteboard. Someone had drawn a caricature of Sojourner, with a smile on its face and its APXS "tail" held straight up in the air, wagging. There were four words above the drawing: "WE HAVE A MISSION!"

TWENTY-ONE

SO WHAT ARE YOU GOING TO DO FOR THE NEXT SIX MONTHS?

"**They're having fun!**" The dark shapes were definitely catching that wave. I concentrated, focused on them in an attempt to capture it on film. I missed it the first time, snapping the shutter a second too late. But more were swimming farther out. I was patient. Some waves rolled in, but they weren't right. No novices these. Finally, the conditions were ideal, and two new swimmers rode the wave perfectly. Click. Sea lions surfing at sunset.

Eight of us sat on the rocky beach of North Seymour Island feeling nearly as carefree as the marine mammals appeared to be. I was in the Galápagos, six thousand miles from home and work, too isolated to hear any news or solve any rover crisis should it arise. There was no telephone and no pager. I was obeying the maxim "Never worry about things you cannot change."

Each morning our boat sailed to a new island landing site and we rode the panga into shore. From there we hiked around, having to watch our feet to avoid stepping on animals that, through isolation for millions of years, had forgotten to fear man. Seabirds—blue-footed boobies—performed their courting dance ten feet away. Land and sea iguanas basked in the sun, looking more like rock formations than living beings. In the hot equatorial midday hours we snorkeled, keeping cool and observing even more creatures apparently oblivious to our presence. In the late afternoon, we went ashore at a new locale and explored further. Each night, the boat's cook produced from the tiny galley prodigious meals with flavors we all loved but sensed we could never duplicate.

A life of leisure was what most of my friends and relatives imagined I could live for the six months before Mars: vacationing, traveling, and just plain lazing about. Pathfinder had been launched; the hardware was on its way to Mars . . . What could I possibly need to do until it landed? Friends knew how hard I had worked, and for how long. Certainly there was plenty of time for a well-deserved break.

The reality would be ten days. Then, after this one sanity-restoring trip to the equator, I would return to JPL to make sure we were ready. We had built the rover, tested its software, launched it. Now we would have to learn to understand the nature of the beast. The rover team was like a teenager who has just passed his driver's test. We had our license, but were we good drivers?

※

On the rover side, I was the one focusing on how we would handle operations. During the integration and test phase of the previous two years, I had begun to sound like a broken record with my repeated warnings regarding the need for operational and performance testing, and I had been forced to defer to the hardware issues requiring resolution. But now was my time. Jake Matijevic had made me the rover mission operations engineer, and my job was to build the operations team. Most of the system team, plus the subsystem Cognizant Engineers, were about to become the core of this new team. But many of those same people had pictured the weeks following launch primarily as a time to relax. We had all come close to burnout in the previous four years. Now, a month after launch, we were in a new phase of the project. The rover team had to move from a sense of a job well done to an attitude of a new job to be done well. I was afraid that the team did not yet have a feel for just how close Mars was, or how much we all had yet to learn. So on January 7, 1997, six months before landing, we kicked off the training of the operations team.

✻

What were operations going to look like? Once Sojourner was on Mars, the rover team would have three things to do, day-in and day-out: analyze the data coming back telling us what the rover had done, write the command sequences telling it what to do next, and coordinate with the lander team. The rover team would divide in three to get these jobs done.

The job of assessing the health of Sojourner would go to the Engineering Analysis team, composed of members from each rover subsystem. This downlink team would be led by the Data Controller, who also had responsibility for massaging the rover telemetry stream into a form the rest of the team could evaluate, and for documenting the results of the team's analysis.

The Rover Coordinator would represent the rover operations team to the project. In the Mission Support Area (MSA), each key lander subsystem was allocated one computer workstation, each marked by a sign hanging from the ceiling above: Power, Propulsion, Navigation, Flight Director. Over another workstation would be a sign marked "Rover." The Rover Coordinator would relay requests from the Flight Director to the rover team, and report key rover status information back to those in the MSA.

The rover's uplink team would be two engineers charged with transforming the desires of the science team into commands to the rover. The Rover Driver would peer intently into the stereo display of the Rover Control Workstation, assessing the IMP images to see where it was safe for the rover to go, and generate the specific commands to get the rover to a selected target. In the meantime, the Sequence Planner would build the rest of the command sequence, including imaging, experiments, and engineering "housekeeping" functions, merging in the traverse commands when ready. The uplink team would submit the final sequence to the lander team for transmission, and document what they had done.

The schedule for all of this would not follow a nine-to-five workday, or any other kind of Earth day. Instead, mission operations would be driven by the Martian day—called a "sol"—twenty-four hours and thirty-nine minutes long. Since Sojourner's day was tied to the rise of the Martian sun, so was ours, Earth-bound though we humans were. The operations team would live on "Mars time": Every day, team members would arrive for work forty minutes later than the day before.

After landing, the process would look something like this: Late afternoon each sol, the lander would downlink to Earth the rover's telemetry that included both engineering data and science results, generated during the rover's activities earlier that sol. Meanwhile, the lander's own telemetry would provide stereo IMP images of where we *expected* the rover to be at the end of its sol's worth of traverse. The rover downlink guys would inspect those images. This would tell them how close the rover had come to its intended target. The downlink team would compare the received rover data with the plan as embodied in the command sequence. Had the rover executed all of its commands properly? Were there any error messages? Was the rover getting too hot? Too cold? Were the rover radios performing as expected?

The analysis team would have about three hours to interpret the day's worth of telemetry and diagnose the health of the rover. The idea was to identify whatever the uplink team would need to know before building the next sol's command sequence.

The entire rover operations team would crowd together in the analysis area for a crossover meeting. The downlink team would tell the uplink engineers and the Coordinator what they had discovered. For the uplink team it was "morning" no matter the actual time: They had just come on shift and needed to be brought up to speed. After the meeting, the downlink team would go off to put together a web page reporting the results of their analysis. Anyone on the project would then have immediate access to the information.

Scheduled soon after the crossover meeting would be the Experiment Operations Working Group meeting. Here, the Pathfinder scientists reported their latest findings and planned the next sol's activities for all of the science instruments, including the rover. When the meeting got around to the rover, the Data Controller would present a brief summary of the rover's status. Jake Matijevic, as Rover Manager, would describe a preliminary plan for the rover's operations. Working Group members would propose changes to the plan, adding experiments and picking new rocks as targets for the rover. At least one member of the rover uplink team would always be present at the meeting to ensure that the rover scenario we were about to be charged with implementing could actually be done. This would sometimes mean saying no to the scientists at the meeting, if what they proposed was too ambitious or too dangerous. One of the engineers' responsibilities was to ensure that the rover lived to do science another day.

With the rover scenario in place, the uplink team would go to work. While Matijevic formally wrote up the scenario as a list of activities, the Rover Driver and Sequence Planner started construction of the command sequence. The sequence would tell the rover what to do for an entire Martian sol; building it would take hours. During the Working Group meeting, the scientists might have come up with ten or so activities for the rover on the next sol; those ten activities would translate into hundreds of commands for the rover. The uplink team had to figure out where to point the IMP camera to get the "end-of-sol" snapshots of the rover. The images would be used to update the rover's position as the starting point for planning the following sol's traverse. We needed to get the camera pointing information to the lander team members sequencing the IMP camera soon enough to give them time to build the picture-taking into their own sequences. When the rover commands were complete, it was the Sequence Planner's turn to write a web page to document the sequence so the downlink guys would know what to expect when the next sol's telemetry came back from Mars.

The rover uplink team would finish its job and go home before the rover ever woke up to the new sol. The drivers of the rover would sleep while the rover roamed, separated by both time and space.

As the Earth rose in the Martian sky, commands destined for lander and rover would flash across interplanetary space from the Deep Space Network station at Goldstone, Canberra, or Madrid. In the early morning, the rover would awaken, and find the lander waiting with new instructions for the sol.

For most deep space missions, engineers built command sequences over days or weeks. On Pathfinder, we would have to do it every day, in about seventeen hours. The whole set of events had to stick to the timetable. A delay early in the process could leave the uplink team too little time to have a sequence ready for the morning transmission, or worse, might introduce errors that could put the rover at risk. Getting behind schedule would result in lost sols on Mars, idle time during which no useful science would be done. No one wanted to waste any of the precious days on the surface, because there was no way to know which sol would be our last.

So when landing day came, we would be operating the rover seven days a week for as long as it and the lander survived. Losing days of activity on the Martian surface so we could get a weekend off just was not going to be acceptable. Yet if we worked seven days a week without a break, we would begin to make mistakes, let alone hate our jobs and become strangers to our families. And those mistakes might well hasten that eventual failure that brought the mission to an end. So each team member would work four long days a week, with two days off. On sol 1, Sojourner's first day on Mars, the entire rover team would be present. There was no way to force people to go home. No one was going to miss the culmination of years of work.

✳

How was the rover team going to get ready for landed operations? We were going to practice. The Pathfinder project had a plan to do Operations Readiness Tests, or ORTs, which were a cross between rehearsals and war games. But they had not scheduled enough of them to suit me. Their first ORT involving the rover wasn't until April, three months before landing! So I proposed we conduct our own Rover ORTs—RORTs—once a week until landing. That sounded like a lot of testing (too much, to many of those on the team), but that was deceptive. We wanted to cross-train people for more than one job so that we'd have some flexibility in case somebody got sick at a critical moment. Between that and the necessity of training enough people to cover seven days a week, each engineer would only get about four tries at his or her job before having to do it for real on July 4.

✳

By early February, it had become obvious that the team was just too thin to cover all of the engineering positions during surface operations. The overall mission operations process would go on pretty much twenty-four hours a day, and some of us were going to have to work very long shifts. While we were all willing to stick it out if necessary; the big question was how good a job we would do. My greatest concern was that the uplink planners would be too fatigued after working twelve hours to pick up on their own mistakes. We needed someone who could come in fresh and review the commands for correctness.

We needed more people. Matt Wallace, from the rover power subsystem, seemed to have just the right temperament to be a Rover Coordinator. He was now dealing exclusively with lander power issues—could we steal him back? I also thought of Sharon Laubach. She was a Caltech Ph.D. student in robotics, working on a rover research task at JPL. She had gotten that job by hanging around the MarsYard helping out until she had wrangled a position doing her dissertation research on the newest Rocky microrover. Laubach had offered to support Pathfinder rover on several occasions. She was bright, she knew rovers, and she was willing. Maybe we could train her to review sequences in time?

I gave Laubach a description of the rover's command set to study, along with a set of command sequences I had generated over the previous several months of testing. I didn't have much time, so I put her in a sink-or-swim position. I was skeptical that anyone could learn the intricacies of commanding the rover in the few months remaining before landing. As Laubach studied the materials, she would formulate questions, then come back to me for specific answers. For a "final exam" I presented her with a complex rover command sequence I had written, and asked that she find the bug in it. It was a particularly difficult bug to locate, for it was not something I had put in by mistake; instead, it was something I had left out. Laubach found the error on her own. She would do just fine!

Stone and I invited Rick Welch to join the team. Welch had been working on rovers of one type or another at JPL for years. He had taken over as task manager on the Hazbot hazardous response robot job when Stone got pulled into Pathfinder. He had since moved on to the Rocky 7 microrover research tasks that had grown up to fill the gap when the first group of rover researchers joined the Sojourner team. Welch was an excellent systems engineer with a background in mechanical engineering. He also had a sometimes vexing sense of humor. One of his favorite pastimes was starting unfounded rumors and watching them circulate.

When we met with Welch, he was low-key but agreed to become part of rover operations. We focused on how to get him involved while allowing him to fulfill his commitments on his current task. Privately, Welch was more excited. As he commented much later: "You guys ask me to drive a rover on Mars. And I'm going to say 'No. I don't think so'?" He stifled a laugh. "Yeah. Right. I'm going to turn down an opportunity like that. Come on!"

One day in the rover downlink team cubicle, Allen Sirota took me aside. He warned me that I was pushing people too hard on operations training. Some of the team members were getting pissed off, because whatever they did wasn't good enough for me. Or at least that was the impression I was giving them. For the past three years, I had resented Howard Eisen's attitude that he owned the rover, that no one else on the team was as responsible for its success as he was. Now, Sirota's few words forced me to confront the possibility that I was guilty of a similar attitude. I could see it in myself: During development, I had been the operations "voice in the wilderness" for so long that today, when the entire team was focused solely on operations, I still acted as if I was fighting the war alone. I was feeling too responsible for the success of the mission. And that left no room for the rest of the team to take on its own responsibility.

Sirota summed up his advice to me in two words: "Lighten up." I would struggle with that suggestion for months.

The Rover Operations Readiness Tests continued. We would never manage to maintain the once-a-week schedule I had hoped for. Even so, it felt as if we were testing Marie Curie all the time. At first, the team stumbled over the mechanics of running the tests: who was doing what, how to interpret the evolving telemetry displays, what information to communicate between the downlink engineers and the uplink team. But the rover team learned quickly.

Even as the team members were adjusting to their new roles for operations, we were having trouble getting the rover to its targets. Marie Curie was just not doing well at going where we were asking it to go. If we commanded the rover to turn twenty degrees to the right, it might turn about twenty degrees. Or it might turn twenty-five. And if that turn were followed by a long traverse of several meters, the rover would drift farther and farther off course as it went.

The onboard turn rate sensor, the rover's primary means of knowing how far it had turned despite the inevitable slippage of its wheels in soft sand, was "drifting." It was not performing consistently. We had seen the problem before, during our abbreviated driving tests with both Sojourner and Marie Curie prior to launch. In the rush to delivery, there had been no time to fully understand the rate sensor drift. Sojourner's sensor had seemed worse than Marie Curie's, so the units had been swapped before the flight rover shipped to the Cape. Other than that attempt to mitigate the problem, we had been trapped by one of the primary constraints of flight projects: The planets don't wait. We had been forced to live with a known hardware problem, for there was no time to correct it. We thought now that we knew the culprit: noise. Just as electronic noise had been the bane of the APXS, noise in the power supply lines feeding the rate sensor was corrupting its output. All we could do was look for software fixes and operations workarounds to somehow compensate for it. This was the nature of one-of-a-kind flight systems versus mass-produced products. An automobile manufacturer would build a prototype, learn about its quirks, and then correct them in the production model. In a flight project, we flew the quirks.

Most of the RORTs were conducted in the Building 107 sandbox, with commands coming from Rover Control Workstation in the flight rover control room on the other side of JPL. But for a few days in March, Marie Curie was not in the sandbox, but instead in the Mojave Desert near an extinct volcanic crater called Amboy. We were finally doing the field tests that Donna Shirley had been insisting on for the last two years.

Weather had aborted the field tests twice before. High winds in the desert along the approach route and at Amboy Crater could make getting to the field site dangerous. Art Thompson had been checking the weather reports daily, waiting for a forecast of at least two days of calm conditions before sending out the caravan of two rented RVs, a NASA flatbed truck, and a van. The motor homes would serve as sleeping quarters and the field command center with workstations and equipment. The truck carried the satellite communications dish antenna and the old Mars Science Microrover (MSM) simulated lander. Marie Curie, safely boxed in its "sarcophagus," rode in the van. The field team consisted of four people: Art Thompson, Allen Sirota, George Alahuzos, and Jim Lloyd. Thompson gave the go-ahead early on a Sunday morning. Their destination was an isolated spot. "We took Highway Forty east, and got off where there's absolutely no civilization, drove for about an hour out to this dead volcano." The farther the small convoy drove up the undeveloped road toward the crater, the more the RVs were sinking into the sand. Once they reached the site, they immediately buried one of the motor homes in the sand up to its axles. It would take a tow truck from the nearest town—over an hour away—to free the motor home for the return trip.

The field team had the test site set up by early afternoon on Sunday, sooner than they had expected. They had set up the antenna dish and established the satellite link to JPL in about thirty minutes. The plan was to get the first stereo images taken by the MSM lander before they lost the light in the late afternoon. Then the rover drivers could plan Marie Curie's first traverse for early Monday morning.

For the field test, the rest of the rover team stayed at JPL. After all, the point of the test was to operate the rover as realistically as possible, with the operations team relying solely on the same kind of data that would come back in telemetry from Mars. So the JPL crew would look only at images and engineering data generated onboard Marie Curie.

It was only a test, but the pressure was on. We needed to get in as much testing as possible while the weather held. Fierce winds could blow the satellite dish over. The sand those winds would carry could scour the outside of Marie Curie, scratch camera lenses, and work its way into the electronics. So we tried to cram three or four sols' worth of operations into each day of testing. The command sequences were dominated by traverses and APXS placement, without any time-consuming experiments to slow things down. The hope was to design the next sequence and up-link it within two hours, let the rover operate for an hour or so, see the re-sults, and start planning another sequence immediately. We'd repeat the process until the lander images were too dark to see what was going on.

The field team spent most of its time waiting. One or two of them would always be there keeping watch over Marie Curie, just out of view of the rover cameras. When the rover was in motion, a field team mem-ber would pick up Marie Curie's power cable and walk along behind, mak-ing sure that the cable didn't get tangled among the rocks. But for hours at a time, the rover did very little. During these periods, the operations team at JPL would be furiously arguing over the best course of action, or building command sequences. The field team had nothing to do but listen to the JPL chatter . . . and plan dinner.

One of Thompson's most vivid memories of the field test would be the meals. They certainly had plenty of time for cooking. "George is an excellent chef. I would say that was some of the best eating we'd ever done." For their efforts Henry Stone had "bribed" the field team with food: He had given them almost five pounds of rock shrimp he'd brought back from Florida in an ice chest after the Pathfinder launch. While the field team gladly accepted the shrimp, they treated it as only the jumping-off point for their own culinary plans. "We ate really well."

The rover testing was not so memorable. There were few rocks in the lander panorama big enough to be good APXS targets. The best target would have been a large, flat-sided boulder, the Martian equivalent of "the broadside of a barn," something so big we couldn't miss it. But there weren't any of those. Instead, the rocks were so small that Marie Curie might accidentally drive right over them. And that meant the APXS would have to be precisely placed to make contact. The JPL crew picked a target. First we had to drive the rover to the vicinity of the rock. Brian Cooper planned the traverses, studying the scene through his stereo goggles. We sent the sequence. When the new images came back, Marie Curie had overshot. We struggled through the next sequences, zeroing in on our rock. When we thought the rover was properly aligned, we commanded

Marie Curie to back up a few inches, then activate the APXS Deployment Mechanism to finally place the sensor head onto the rock.

The new images were a disappointment. The APXS was only inches away from its target, but we had missed again. I was frustrated, and taking it personally. I knew we could hit the rock with one more sequence. Perhaps the MSM lander cameras had been knocked out of calibration during transport to the site. Perhaps the rate sensor drift was throwing off the rover turns. But we were out of time.

By noon Wednesday a new weather front was moving in. The field team came home. The small group that had braved the desert boredom had now been jokingly branded "The Amboys."

The mixed results from our attempts to get to the target rock at Amboy Crater, together with similar experiences during earlier sandbox RORTs, left the team with a looming uncertainty: Would we really be able to reach rocks in a day or so on Mars? Would we be able to make Sojourner's mission a success?

We would keep practicing.

TWENTY-TWO

TESTING, TESTING . . .

Friction between the rover team and the project team had persisted for most of the lifetime of the project. To many of the rover team members, the dichotomy was simple: "We" were the rover and "they" were the lander.

The difficulties between the two teams were most evident during the integration and test phase of the project. Everybody was working hard. Everyone was under pressure. To the rover team, the lander team seemed unresponsive. To the lander people, the rover group was arrogant and demanding. The lander had its own problems to deal with; they'd get around to things the rover needed, but the rover was never willing to compromise. The rover guys always wanted their tests run *now!*

The rover and lander engineers saw themselves as two separate teams. But after launch, as the Pathfinder spacecraft moved inexorably toward its rendezvous with Mars, that perception had to change. As lander system engineer Dave Gruel recalled later, "What really made us one team were the ORTs [Operations Readiness Tests]. After the first one, we realized we had a common enemy. We were all flailing around. No more was it 'My stuff works and yours doesn't!' Now it was 'My stuff's not working. Your stuff's not working. We've got to come together and make it all work, or we're in deep trouble.'"

The "common enemy" paradigm had come into play throughout the design and development of the rover and Pathfinder, operating at several levels. Conflicts between rover subsystems tended to strengthen the camaraderie among the individuals within a subsystem. During the protracted effort to solve the APXS noise problem, Tom Economou, the "Principal Investigator from Hell," became the bigger enemy, and thereby helped the rover team to bond more tightly together. The rover team was further unified by our shared perception that the lander agenda threatened the ultimate goal of a successful rover mission. And then, with the prospect of mission failure looming ahead, as simulated in the Operations Readiness Tests, the rover and project teams finally began to meld together.

The ORTs were full-up simulations of on-Mars operations. In order to make those simulations seem as real as possible, the project needed its very own little piece of Mars: the "sandbox." Just as the Building 107 sandbox had been the rover team's arena for testing and improving the rover's performance, the Pathfinder sandbox would validate overall surface operations. The project's sandbox was a sealed room on the second floor of the Space Flight Operations Center. White sand six inches deep covered the floor. There was a small anteroom in the corner through which people came and went; this required anyone entering to pass through two sets of doors. There were two reasons for this: first, to protect people from the limited but real hazard of the rover's laser stripe projectors, and second, to prevent the migration of sand from the sandbox to the rest of the second floor. The mission operations area was filled with Macintosh and Sun workstations, and sand was the bane of computer hard drives. The engineering model of the lander was placed in the sandbox. Track lighting allowed for sufficient illumination to operate the IMP and rover cameras, but not nearly enough to simulate Mars sunlight. A photomural of one of the Viking landing sites covered part of two walls. The flight testbed was located on the other side of the sandbox's north wall. A window in the center of the wall allowed the engineers in the testbed to observe the activities in the sandbox. There were also two windows on the south wall of the sandbox, so people in the adjacent hallway could look in. During tests, blinds covered those windows. The doors to both the testbed and the sandbox were sealed by cipher-locks. Only those people with the right combination could get in. The point, during ORTs at least, was to make the sandbox as far away as Mars. The operations teams would only know what they could discern from the telemetry stream coming back from the lander and rover. The teams would have to learn how to command the mission, assess the health of the spacecraft, and recover from anomalies with only the incomplete information Pathfinder could send home.

As part of giving the operations teams much needed experience operating the lander and rover the way they would have to on Mars, the readiness tests forced the teams to live the work shifts that would be in place after Pathfinder landed. We would be generating new commands while the lander and rover "slept" during the Martian night. Only that way would commands be ready to transmit to the lander when Earth rose in the early morning Martian sky and the big antennas of the Deep Space Network had direct line-of-sight to the lander.

And it so happened that on July 4, the time at the landing site would be about seven hours earlier than Pacific Daylight Time on Earth: The landing would take place about 3 A.M. Mars Local Solar Time (MLST), which would be about 10 A.M. in Pasadena. I was going to be the Sequence Planner for sol 2. Two days before each ORT, I would force myself to stay up as late as I could in an attempt to shift my personal biological clock to the midnight shift. I would rent four or five videos, then struggle to remain awake until the sun came up.

⁕

The ORTs were always focused on landing and the first few sols of the surface mission, or some subset of that time period. The activities of the early surface mission required the execution of a complex choreography to put the lander and rover in a stable state so that they would survive and achieve the mission objectives.

Rob Manning and part of the lander operations team simulated Entry, Descent, and Landing (EDL) over and over again. In some sense there was not much in the way of EDL "operations" for the team to train for: the lander acrobatics necessary to drop from interplanetary speeds to a stable perch on the surface of Mars would happen too fast for any human intervention from Earth. EDL would begin and end in less time than it would take any message from the lander to arrive at the DSN. In fact, the entire Entry, Descent, and Landing process would be activated by a single command from the Pathfinder operators, called "Do_EDL." And this command merely authorized the lander to begin running the onboard software that would do the right things when the time came.

Despite all of that, there was great value in running simulations of EDL. There might be subtle bugs still hidden in the software. Slight anomalies or perturbations to the lander's performance might affect the software's execution in unexpected ways, perhaps causing EDL to fail. If Manning and his team could find any such sensitivities in simulation, there was still time to correct them, and upload new software to the Pathfinder spacecraft already en route to Mars.

Simulating EDL would also make the next part of the ORT more realistic. The EDL telemetry data would be some of the first information received from the spacecraft after a successful landing.

The operations teams in the ORTs would attempt to follow the detailed sol 1 scenario that had evolved over the prior three years of the project. For a successful start to the surface mission, the series of events in the scenario would need to occur in rapid fire. As long as there was no damage to the spacecraft due to the landing itself, most of the steps that had to happen on sol 1 could be handled by a set of pre-canned sequences. The lander and rover operations teams on Earth would just be required to make a number of go/no-go decisions, then send the command to activate the appropriate sequence already stored in the lander's memory on Mars. After sol 1, the operations team would normally command Pathfinder only once per sol. While this would be sufficient for most of the mission (and represented a much faster command turnaround than in any prior mission), it would not be good enough for sol 1.

Once the lander was on the ground, it would transmit a confirming signal through its thumb-sized antenna. That signal would not be heard on Earth, unless the lander happened to come to rest base-petal down, so that its tiny antenna pointed at the sky. But that was the best the lander could do for the moment. It had other tasks to perform before it would be

able to send a more certain message home. The airbags would need to be
vented and retracted out of the way; otherwise they would remain draped
over everything near the lander, precluding close-up images of the local
terrain by the IMP camera, and impeding rover egress. So retraction motors in each lander petal would operate, reeling in the Kevlar cords that
ran through the airbags onto take-up spools, dragging the bags across
the ground, and compressing them more tightly against the backside of
each petal.

Airbags safely stowed, Pathfinder would proceed to open its petals,
the inside surfaces of which were mostly covered with solar cells to re-
plenish the depleted lander battery. Only the central base petal, to which
the other three were attached, was free of solar cells; instead it carried the
lander's computer, radio, antennas, and instruments. In the act of petal
opening, the lander would set itself upright, no matter its original orientation. Onboard sensors would tell the lander how it sat on the surface.
Whichever petal was down would open first, until Pathfinder fell onto its
base petal. Then all three outer petals would open fully, leaving Pathfinder
ready for its first Martian sunrise.

With the petals open, the lander's low-gain antenna would now be upright and exposed to the heavens. Less than four hours after landing, and
a little over an hour after sunrise, Pathfinder would begin transmitting the
first message likely to be detected on Earth by the DSN. The contents of
the message would be engineering data about EDL, along with status
about the lander and rover subsystems. But the meaning would be "I'M
HERE!"

The operations teams would now have some work to do, quickly reviewing the downlinked telemetry. If no significant anomalies were found,
they would then uplink commands authorizing the lander to continue
with the nominal mission plan. Still in its stowed position, the IMP camera would begin taking stereo pictures. When combined, these images
would form the "insurance panorama," so named because it would ensure
that at least some photographic record of the landing site would make it
to Earth in case the IMP failed soon after landing. (One clearly identified
potential glitch that could cause the loss of the IMP would be a failed mast
deployment. The IMP mast was a spring-driven device with significant

energy stored in it, similar to the rover ramps. Stowed, it was only about two and a half inches tall; when released, it snapped upward to its full height of over three feet. Locked in its deployed state, the mast was a rigid structure, just what you wanted to hold a camera steady for picture-taking. But if the lander happened to land on a rock that tilted its base more than thirty degrees, the IMP mast might deploy unevenly and fall over. The mast was like a sophisticated version of a jack-in-the-box. If the deployment failed, the IMP head would end up hanging upside-down, useless, bobbing in the breeze. There would be no second chances. There was nobody to put the jester back in the box to try again.)

With the sun high enough in the Martian sky, the IMP camera would begin searching for that sun. Software running on the lander's computer would look for a set of bright pixels in the images. When found, the software used the position of the sun in the sky, together with the known time of day, to uniquely determine the direction of north. (Mars has no magnetic field, so a magnetic compass would not function there.) Using its clock to know the time and knowing its orientation on the Martian surface, the lander could now track the most important object in its sky: the Earth. The lander would point its high-gain antenna and at the proper time begin transmitting data, including images, at a far greater rate than the low-gain antenna could ever achieve. If everything were operating perfectly, the transmission would start at 9 A.M. Mars local time, about six hours after landing.

Among the downlinked photos would be those needed by the rover team to decide if one or both ramps could be safely released. The Rover Driver and other team members would examine the images, in stereo, looking for any rocks or rises that might block the unrolling ramps. If the rover team determined that it was safe to deploy the ramps, a new set of commands would go up to the spacecraft. More pyros would fire, and the tie-downs of both the rover and its APXS sensor head would be cut loose. The ramps could not yet be released: The position of the APXS sensor head put it close to the rear ramp, which might strike the APXS at the moment the ramp unrolled. The next rover command sequence would cause the rover to retract the ADM fully, taking the APXS out of harm's way. Moments later, the ramps would deploy.

The third rover command sequence of sol 1 would tell the rover to stand up. The rover would drive its rear wheels forward, rising to its full height and maximum ground clearance. From this vantage, the rover could then take the first set of images with its own cameras.

The IMP would take more pictures, this time to confirm that the rover ramps were properly in place, ready to be driven on. After the unavoidable time delay caused by Mars's great distance, the images would arrive on Earth, and the rover operations team would scramble to answer the questions, "Which way do we drive? Forward or back? Which way should we go at the bottom of the ramp?" From this point on, no completely pre-canned rover sequence would do. The terrain at the end of the ramp would be totally unknown until those first IMP images trickled in. The Experiment Operations Working Group would now recommend the first APXS target, either soil or a convenient rock. The driver would then plan the rover's first traverse.

Dozens of pyrotechnic firings were necessary to get Pathfinder safely to the surface: cruise stage and heatshield separations, parachute deployment, airbag inflation, and more. Each of these pyrotechnics was triggered using energy from the lander battery, and the combination of all of them would have left the battery severely depleted. As the sun rose on sol 1, the lander battery would likely have less than half of its capacity remaining. Sunlight on the solar panels would begin recharging the battery, but as long as the rover sat on one petal, most of that petal's solar cells would be shadowed, making the petal useless as a power source. The lander's total solar power would therefore be down a third. The charging of the battery might not keep pace with the drain from the various lander subsystems. The lander operators wanted the rover off the petal as soon as possible, to bring the solar arrays up to full strength.

Following the sol 1 plan, the rover team obliges. The new sequence is uplinked, and the rover drives. The rover rolls off the ramp and onto Mars. The IMP takes an end-of-sol photo of the predicted location of the rover. When the images finally arrive on Earth, they confirm that the rover had done its job. Toward the end of the first sol, the lander operations team commands the lander to deploy the IMP mast to its full height.

The lander and rover would now be ready to do science.

And any ORT that reached this point in the plan by the end of sol 1 would be going smoothly indeed.

❋

Dave Gruel was the "Gremlin," the engineer responsible for injecting problems into the ORTs. He would rearrange the terrain in the sandbox, creating dunes and rock distributions as he saw fit, positioning the lander to challenge the operations teams. The job required creativity, a bit of a mischievous attitude, and the ability to keep a secret. Gruel was good at it. The final job qualification for a Gremlin was a thick skin, since people's tempers often flared at the person responsible for making their lives so difficult. Gruel could handle that part of the job too.

For the first full-scale surface operations ORT, Mission Manager Richard Cook had told Gruel to make the terrain difficult. It would be the first dose of reality for the team. Gruel obliged. He placed the lander on a large dune, which he and Rob Manning christened "Olympus Mons" after the Martian volcano that was the largest such feature known in the solar system. The problem for the rover team would be deciding which ramp, if any, would allow the Marie Curie rover safe egress. Gruel designed the terrain carefully, to create a situation that was challenging but doable. There was a trough between where the lander sat on Olympus Mons and the next rise in the sand where the forward ramp would rest if deployed. That meant that the ramp would have to act as a bridge. To make sure the rover team chose the forward direction, Gruel saw to it that a deployed rear ramp would slope steeply downhill, at a forty-five- or fifty-degree angle, beyond what the rover had been designed to negotiate. "I expected the rover guys to say it was way too steep to go off the back. I purposely made the tilt horrific."

But things did not work out according to the Gremlin's plan. When the first images came back after "landing," Howard Eisen was furious. As he saw it, the high dune and trough next to it made rover drive-off nearly impossible. It was a totally unrealistic situation! He was convinced that such terrain would never actually happen on Mars. He went to complain to Gruel. When Brian Cooper estimated from IMP stereo images where the forward ramp would contact the ground, it was obvious to him that it was going to be unacceptable. There would be no room for the rover to maneuver once it drove off. Therefore, the rover had to go down the rear ramp! The rover team requested that only the rear ramp be deployed.

The few people who were allowed into the sandbox during the ORT, like Dave Gruel and Jake Matijevic, were precluded from participating in the key operations decisions taking place. It was very frustrating to them, because they could only sit by as the rover operations team made choices based on the telemetry data alone, and could not so much as exhibit a revealing facial expression.

For the next three days of the ORT the rover and lander teams worked together to command the lander petal deployment motors to tilt the lander into a better position for deploying the rear ramp. A group of scientists were participating in the test, waiting for the chance to practice directing the rover to science targets as they would when Sojourner reached Mars. But as the ORT proceeded, it seemed unlikely that the rover would be going anywhere, and from a science training perspective, the test looked like it was going to be a washout. But, finally, the rover operations team was satisfied and asked the lander team to deploy the rear ramp.

The project could not afford a perfect duplicate of the flight lander to complement the flightlike Marie Curie rover. Instead, the sandbox lander achieved flightlike capability piecemeal. The lander's computer was actually located in the next room, in the testbed area, with cables running through the wall to the lander hardware in the sandbox. When it was time to deploy the IMP camera, someone would bolt a mast of the appropriate height to the lander body, and the camera itself to the top of the mast. "Deployment" of a rover egress ramp consisted of an engineer manually setting down and aligning the ramp with one end on the lander petal and the other on sand.

Although Matijevic was the Rover Manager, during the ORT he took on the role of carrying cables, making sure the Marie Curie did not drive over its own power lines, and deploying ramps. When he heard that it was time to deploy the ramps, he placed both engineering model ramps in place, one forward and one rear. He knew from his direct view of the rear ramp that the rover could never drive safely down such a steep angle. Clearly the forward ramp was the way to go. It was only later, after the IMP camera had taken the stereo rover deployment panoramas, that Matijevic discovered the rover operators had not intended for the forward ramp to be deployed. He rushed over to Rover Control to intercept the images before the operators looked them over. The IMP panoramas would have to be taken again, without the erroneously positioned forward ramp in place. For better or worse, the rover operators would have to live with their decision. Relying only on appropriate telemetry, it would take the rover team another full day to conclude that the forward direction was the only viable solution.

On sol 3 Marie Curie finally drove down the forward ramp. Some APXS data was collected, and some images were captured. The ORT ended ingloriously on sol 5.

At the end of the ORT, Dave Gruel hated his job, the rover team, and the scientists. He had done the best he could to challenge the operations team, which was exactly what he had been asked to do. But everyone had come screaming to him to complain. The rover team had complained that he had made their job impossible. The science team had pointed out that Gruel was not a geologist, and that the terrain features he had created could not exist in nature. He had gotten an earful. From the operations teams' point of view, nothing had gone right. Looking back at it, Gruel considered that ORT "the biggest mess, you could not imagine a test going worse."

Yet, from another perspective, the ORT had been a major success: It had proven that we were not yet ready for Mars. Neither the operations team, its procedures, nor all of its software tools were yet mature. The test had revealed problems early, so that they would not happen again when the actual landing occurred. It also marked the end of the hardware bias of the Pathfinder project. Surface operations preparation had necessarily been on the back burner during the rush to get the spacecraft built, tested, and launched. Now everyone on the project had had an experience that told them operations would not just take care of themselves. And for most Pathfinder project personnel, the ORT had been the first exposure to the steps involved in operating a rover. The Sojourner and Pathfinder teams began to grasp the complexities of operating their respective spacecraft.

※

The ORTs slowly improved. The teams calmed down between the tests, and focused on what they had learned. Dave Gruel continued to invent new problems for both lander and rover teams to overcome. The teams even became less belligerent toward the Gremlin.

To set up the sandbox, Gruel would come in on the Sunday afternoon before an ORT, when no one else was around. He'd go into the sandbox with a shovel and rearrange things. If he didn't like the results, he'd rearrange the terrain again. While the rest of his job was getting better, he still hated setting up the sandbox. It was exhausting physical work, and shoveling kicked up clouds of fine dust that he had to try to avoid breathing. Partly to make the setup task less distasteful, and partly because his "victims" could use a break, Gruel started to place foreign objects into the sandbox before each test. He couldn't make the ORTs any easier, but he could find ways to inject some humor into the tests. "ORTs were a high-stress time. If there was something we could do to lighten things up, that was a goal." For one ORT, Gruel positioned a plastic skeleton on a rock, facing the lander and gesturing toward the IMP. In another test, a child's sand pail and plastic rocket toy were visible. After a few ORTs, the participating scientists and the operations teams were in the habit of searching the downlinked IMP images for the Gremlin's signs, and wondering whether they would recognize all of them.

The geologists studying the initial "mission success panorama" during one test indeed discovered something unusual. Additional multispectral imagery confirmed the find: a potted plant. At the Experiment Operations Working Group meeting, the science team took the bait, recommending that the rover get a close look at this evidence of potential Martian vegetable life. The rover team was less sanguine about sending the rover to investigate. The leafy green life form—tentatively identified as a philodendron—was in the far corner of the sandbox, and would require a traverse of several meters to reach. The rover never did get to the plant before the completion of the ORT, but we did manage the longest rover traverse up to that time.

For another of the ORTs, Gruel draped deflated airbag material over the rover petal, blocking the deployment of both ramps. The lander team came up with a scheme to solve the problem. They put together a set of commands to lift the offending petal off the sand, operate the airbag retraction motors, drawing the airbag material more tightly against the outside of the petal, then lower the petal back onto the surface. After the sequence went to the testbed, IMP images confirmed that the airbags were out of the way, and the rover mission proceeded.

What if the lander's battery failed after landing, allowing the lander to operate only when the sun was high in the sky? What if the high-gain antenna never locked onto the Earth, and the entire mission had to rely only on the low-gain antenna? The ORTs gave the operations team experience dealing with contingency scenarios like these. The sequence builders came up with new commands, and the engineers monitoring power profiles and thermal conditions learned the use of their software tools, and refined them when necessary. ORTs did not go perfectly, and there was never enough test time for the engineers to gain all the expertise they wanted. But the operations teams learned to be flexible. This was key. As Richard Cook was fond of saying, "No plan will survive sol 1."

The rover team continued to do its own RORTs, interspersed with the project's ORTs. There was a lot we were learning without the constraints of a full-up ORT involving the entire project operations team. At one point Bill Dias, the project's lead surface operations planner, joked that we should conduct Surface Normal ORTs or "SNORTs." We never did.

＊

In early May, the Hot Wheels toy rover hit the stores. Howard Eisen had worked nights and weekends with the Mattel designers over the period of a year to design a tiny model of Sojourner. Though the sets were initially in short supply, we all scrambled to find one.

I went home one Friday, figuring I would call a few toy stores to see if they had any in stock. The first store said simply no. The operator at the next store didn't know if they were in stock, so she had me wait while she connected me with someone on the floor. When I explained what I wanted, I was told, "Oh, yeah, we had a few come in on Wednesday, but only a few. Three cases of three each. They've all been sold." The guy on the phone had no idea when more would come in. I asked where else I might look. There was a pause. "Well, you know, I do have one that I set aside for myself. I don't really collect them, so I could let you have that one." I got his name and headed out to the store, which was all of ten minutes away. Along the way, I considered that he was probably going to try to scalp the rover. I wondered how much I would pay, and how much cash I had with me. When I got to the store, I quickly located the salesman. He took me to the back and brought out the box. Inside was the model of Sojourner, flanked by the Pathfinder lander and the cruise stage. "Is that what you're looking for?" he asked. "That's it." Instead of asking for money, he simply turned and walked away, disappearing down a store aisle. I went up to the register and paid: five dollars plus tax.

A few days later, I passed by Matt Wallace's office. He had a funny look on his face. He looked over at me. "I just got a call from my mother. I mentioned to her a few days ago that the Hot Wheels were out, and that they were hard to come by. She's like a bulldog. She must have called all the toy stores in the Washington, D.C., area. She just called to say she had located a case that she thinks has at least twelve rovers in it. What did I want her to do? I told her, 'Buy it! Ship it!'"

My secretary would report to me her progress on locating the rovers every time I saw her. She was clearly on her own crusade. "There's none available in Arizona!" she announced one day.

Eventually, just before landing day, a large number of toy rovers arrived and were distributed, one to each member of the Pathfinder team—"payment" from Mattel for the design assistance provided by the Sojourner team.

✳

The rover primary mission had always been stated to be one week, or seven Martian sols. But most of the rover team thought the rover would keep driving after that. Matt Wallace said that if the lander and rover survived landing, and we were able to get Sojourner stood up and down the ramp, "the rover will last for a good long time." He was betting the rover would survive as long as the lander.

The attitude of most of the scientists was very different. Many of them saw the rover as a nuisance, draining resources they would rather use themselves. The rover would cut into the available communications bandwidth with the telemetry volume it generated: Every bit of data sent down to Earth to support rover operations would be one less bit available for their science. They took the promised seven-day lifetime of the rover as gospel, and assumed that after one week on the surface, Sojourner would be dead, and they could proceed with their science unencumbered.

Most of the science team also seemed to oversimplify rover operations, imagining that the rover would just trundle from rock to rock on a daily basis, getting a new APXS reading each night. From all of our RORTs and ORTs, the rover team knew that getting the APXS on target was *hard*. Tom Economou's viewpoint was that the rover team just wasn't committed to getting APXS data. At one science planning meeting during an ORT, he stood up and raised his fist into the air. "I challenge the rover team to get to this rock!" he demanded in his raspy voice. Project Scientist Matt Golombek agreed that the rover team should go for broke and try to reach the chosen rock in one attempt. That attempt required a complex set of maneuvers across several meters, which together with Marie Curie's inherent dead reckoning error resulted in a final rover position a long way from the target. Golombek got the point: "We are never going to do that again. We are going to work within the capabilities of the rover. If it takes us two days to get to a rock that's what it's going to be. We're not going to go ten meters to get to a rock. We're going to go a meter or two." But would the rest of the science team understand?

✳

Just before the final ORT, Hank Moore, now the officially appointed Rover Scientist, gave a talk he called "The Cussedness of Inanimate Objects." In his late sixties, Moore was the "old man" of the science team. He had been a planetary geologist working at the United States Geological Survey since the 1950s. In the mid-sixties he helped select the landing sites for the manned Apollo missions to the Moon, and gave the astronauts geological training before they got there. But Mars had always been a special place for Moore; in the 1970s, he identified sites for the Viking landers, and continued analyzing the geology of Mars after the landers touched down. His ongoing study of the planet eventually led him to derive the widely known—at least among planetary geologists—Moore model of rock distributions: This was a mathematical representation of the number of rocks of a given size you'd expect to find at a given location on Mars. (When rover designers at JPL designed test courses, or analyzed terrains to determine what size a rover needed to be to overcome obstacles, they used Moore's model.) In time, Moore retired from the Geological Survey. But he came out of retirement to be the Rover Scientist on Mars Pathfinder. He seemed to be always smiling. And he praised the rover operators whenever they got Marie Curie to do something like it was supposed to do during ORTs. The impression you got was that he was having the time of his life.

Moore's involvement in Viking twenty-two years before gave him a unique perspective, and his presentation had a good turnout. His basic premise was that even rocks wouldn't do what you expect them to do. He warned the Pathfinder team that rover operations would not go as smoothly as they were assuming. During the Viking mission, "We tried to push the rock called ICL (for Initial Computer Load) with the Lander 2 sampler. ICL did not move. Another rock, Badger, rotated and leaned on the Lander 2 surface sampler. Badger finally yielded, but he is still there on his haunches, ready to do battle with the Viking surface sampler." During Viking, it had taken weeks to move a few rocks around. The average delay between when a scientist proposed an experiment on Viking and when the commands finally were sent to Mars had been twenty-two days. On Pathfinder, the rover team was going to be turning around sequences every sol! Moore closed his talk with a plea to the assembled scientists for patience during rover operations. One of the great desires of Pathfinder was to find "fresh" rock, free of dust, for the APXS to analyze. Placed against a dusty rock, the APXS would probably determine the composition of the dust, instead of the rock under it. "Locating such rock surfaces will take time. Demonstrating the accessibility of such rock surfaces will take time. Accessing and positioning the APXS will take time." If Sojourner didn't get to that special rock in a day, well, she would get there the next day, or maybe the day after that . . .

Later, back in the office we shared, Henry Stone and I declared Hank Moore the rover team's savior. He seemed to be the only person on the Pathfinder science team who both had the stature to be listened to by the others and understood our problems. After the many months of arguing with Tom Economou, who seemed to feel that the rover team didn't care about his instrument, it was good to know that there was someone in our corner.

*

The seventh and last ORT was our dress rehearsal for landing day and the week of surface operations that would follow it. For the first time in all of the ORTs, the simulated mission went smoothly. The lander experienced few anomalies, and we got the rover down the ramp on sol 1, just as planned. The terrain of the landing site was relatively flat; the rover maneuvered well in the sandbox and reached its designated targets. The rover team reported Marie Curie's progress each day in the Experiment Operations Working Group meeting, and the science teams set priorities for the next sol's rover activities. The rover took pictures and operated its APXS. All in all, it was a successful one-week mission.

Of course, it was all planned that way. Richard Cook had asked the Gremlin to make this ORT an easy one. It was mid-June and the real Pathfinder landing was only a couple of weeks away. The purpose of the final ORT was to instill confidence in the operations teams that we were ready for Mars. The teams were as trained as they would get. Dave Gruel inserted no significant problems to vex the operations team. So for those of us participating in the ORT, this was our first experience with nominal operations. Compared to the challenges Gruel had thrown at us for the past few months, operating a deep space probe twenty-four hours a day, seven days a week with no disasters was a piece of cake.

*

The second rover cruise healthcheck was scheduled for June 17, exactly six months after the first healthcheck. While we had been conducting ORTs and RORTs, testing and training with the Marie Curie rover and learning to work together with the lander team, Sojourner had been asleep, conserving its batteries, during the long trip to Mars. This "late cruise" health-check was also our chance to load Sojourner with the software updates Jack Morrison and Tam Nguyen had produced since launch.

We loaded two sequences onto the Pathfinder spacecraft. After confirming the loads, the lander team activated the first sequence, which would wake up the rover and give it the forty-four software patches we had approved. The waiting was more excruciating than the first health-check. Back in December, the travel time for a radio signal from the spacecraft was measured in the tens of seconds. Now Pathfinder was so far away that a message traveling at the speed of light took over nine minutes to make the trip. So the minimum wait between sending a command and seeing the result was nearly twenty minutes.

But the results did come in. Sojourner was up and running. The telemetry from each rover command showed that Sojourner was accepting every software patch. After more than ten minutes, the sequence completed and the rover shut down once more. The software changes we had sent would take effect the next time the rover was powered on.

A quick analysis by the rover downlink team indicated no rover anomalies. Sojourner appeared to be as healthy as when she started her voyage.

✳

At the Rover Coordinator's request, the second sequence was activated. This was a typical self-diagnostic sequence, nearly identical to the one we had used six months before, which would confirm that the APXS instrument was still operating properly. Normal execution of the commands would also demonstrate that the software patching of the prior sequence had not broken the onboard software. After the requisite time delay, the telemetry downlink arrived. Sojourner had done what we had asked of it without complaint, and then shut down as planned.

Sojourner's safety was now in the hands of the lander team. The next time we heard from the rover, it would be on the surface of Mars. I tried to suppress the nagging thought that we might never talk to Sojourner again.

TWENTY-THREE

MOMENTS OF TRUTH

n the last few weeks before July 4, I found that I just wasn't hungry, and I began losing weight. We had reached the point where there was nothing to do but worry over whether the landing was going to succeed. After all, there were a lot of things that could go wrong.

No one really knew what the chances of a successful landing would be. EDL was a complex process. For example, there were forty-one individual pyrotechnic events (actually small explosions), every one of which had to happen in its own time during EDL. Each pyro firing would release or separate a part of the spacecraft when appropriate during the descent toward the Martian surface; the failure of any one of these would doom the landing.

One of the key elements of EDL was "aerobraking." This would occur when the lander, protected by its heatshield, entered the Martian atmosphere at about seventeen thousand miles an hour, and then lost 95 percent of its velocity in just a few minutes. Even the thin atmosphere of Mars would provide sufficient drag on such a fast-moving blunt object to decelerate the lander to a speed where the parachute could be safely deployed.

Rob Manning was the Pathfinder flight system engineer, the lead engineer responsible for all aspects of EDL, from design to implementation to operation. In a fit of gallows humor, he considered an alternate braking scheme. Just as an "atmosphere" was the gas layer that enveloped a planet, so a "lithosphere" was the top solid layer of a planet—the surface. Manning coined a new term that would come into play in case any of the pyros failed to fire on time: lithobraking. The surface of Mars would effectively bring the lander to a halt, no matter how fast it was moving when it hit. Of course, there would be little left of the lander . . . Tim Parker, a science team member who had been wondering about how terrain features discovered on Mars by Pathfinder would be named, responded that if lithobraking was used, the only new feature named by Pathfinder would be its impact crater. The rover team chimed in that, after lithobraking, the rover would be transformed into a subsurface explorer vehicle.

Lithobraking started out as a joke, but as landing day approached, our nervousness made it seem less funny.

Outside the Mission Support Area there was a chart covering the wall showing the timeline of operations activities and associated staffing. The timeline indicated when critical spacecraft activities would occur, when commands would be sent, telemetry received, and when individual team members would be on station. The detailed plot of this information showed events happening twenty-four hours a day for several sols. The timeline was updated for each Operations Readiness Test and titled accordingly as "ORT 6," "ORT 7," etc. One day I walked by and saw that the timeline was now labeled simply "FOR REAL."

There would be no more tests.

July 4, 1997

I was watching NASA TV on a monitor in the rover engineering analysis area, about fifty feet from the Mission Support Area. NASA TV was running a live feed from the MSA, focused mostly on Rob Manning. The VOCA voicenet was also turned on, so we were hearing what was said twice: first, over the VOCA, then, a few seconds later, from the TV monitor. If there were cheers or applause, we'd hear them three times, because they were loud enough to be heard directly.

I was nervous as hell, and there wasn't anything for me to do but watch. I wondered whether I'd have a job to do tonight. It all depended on the landing. Rob Manning continued to describe what was going on with the spacecraft. His narration told the world what should already have happened ten minutes earlier; the signal from the spacecraft, if it were detectable, would only just be arriving as he spoke.

It was just after 10 A.M., Pacific Daylight Time. Sharon Laubach, the Caltech graduate student who had trained to review rover command sequences, was sitting next to me as we watched the video feed. I was sure that we both had the same sick expression I saw on her face. When we were about ten minutes from possibly receiving the landing telemetry—which meant Pathfinder was hitting the surface *now*—I turned to Laubach and said, "I'm sure everything is going to be okay." "Is that a question or a premonition?" she asked. "Premonition." For some reason, I had just decided the landing had worked.

And the next few minutes proved it. We had expected to lose the radio signal from the spacecraft at the moment the parachute deployed. Instead, we detected it almost all the way to the surface. A few minutes later, Sami Asmar, the communications engineer who had been flown out to the DSN tracking station in Spain, reported in. "Comm, this is Madrid. I see a weak signal . . ." Pathfinder had survived the landing! Cheers went up on the second floor of Building 230. We had thought we would have to wait hours for confirmation, but the signal came through almost immediately. And the length of the next signal told us that the spacecraft had landed right side up, sitting on its base petal. Rob Manning whooped. "What are the odds!" he bellowed, a huge grin on his face. Of course, the odds were about one in four, since Pathfinder was a four-sided tetrahedron. So far, Pathfinder had been lucky.

Four hours later it was time for the first telemetry session from the lander. If all had gone well, our lander would have retracted the spent airbags that had cushioned its impact on Mars. Then the petals would have opened, revealing the lander's solar panels and exposing the rover itself to an alien sky and the soon-to-rise Martian sun.

"We've got lockup!" The DSN had locked onto the low-gain transmission from the lander. Everyone in the MSA was ecstatic. "We have rover data!" Matt Wallace announced over the voicenet.

The lander had switched on the reed-relay, waking up the rover, at 6:59 Mars Local Solar Time (MLST). Healthcheck data came in from the rover. Some of the data looked wrong. Suddenly one of the engineering analysis team members realized we were seeing symptoms of the same data conversion problem he had seen during ORTs with Marie Curie over the past few months. "The first healthcheck is bad. We can't trust the data. The next one should be okay." The engineering analysis team understood the system well enough to know that the bad initial data was not evidence of a hardware problem on Sojourner. Then the telemetry arrived from a commanded healthcheck at 7:35 A.M. MLST. Sojourner looked healthy!

Lander data was coming in. The tilt of the lander would be critical to deploying the ramps and getting the rover off. If it were too steep, we'd have to adjust the petals to reduce the angle to something the rover could negotiate. One of my nightmares had always been that the landing would go perfectly, but the rover would never be able to drive. The tilt of the lander was . . . two degrees! Pathfinder was sitting virtually level on Mars. We hadn't seen pictures yet, but how much more could go right?

*

The biggest uncertainty for the Pathfinder project had been whether the previously untried technology for Entry, Descent, and Landing would work. With a successful landing, most of the Pathfinder team was thrilled, elated . . . and beginning to relax. Everything else was gravy. The rover team was thrilled, elated, and nervous. No one had really known if the landing would succeed; the uncertainty had stood as a psychological barrier between us and the mission we would perform with the rover. That barrier was now down. The lander hadn't blown its big moment.

Now the rover team's big moment was fast approaching, and we hoped we wouldn't screw it up.

Knowing that Sojourner had weathered the landing, I also knew that I should try to get some sleep. My shift wasn't until that night, and the first images weren't due in until after 6 P.M. I didn't want to leave, but I didn't want to mess up the most important job of my life by being totally exhausted. I borrowed the key to Art Thompson's RV, which he'd parked near the Space Flight Operations Center as a makeshift bunkhouse for rover team members, and forced myself to leave the center of the action for a few hours. I tried to sleep for some time, but just couldn't manage to doze off. Finally, I headed back to the rover control area.

I was in time for the first high-gain communications pass from Mars. Gordon Wood, the lead lander communications engineer, had warned the project months before that the high-gain antenna might never get aligned well enough to transmit a good signal to Earth. If he was right, we'd be stuck with a low-data-rate mission. The signal came in strong, stronger than even our most optimistic plan had predicted. Wood had been wrong, in the best way possible. The IMP had found the sun, the high-gain antenna was pointed straight at the Earth, and the first images were coming down! Pictures from Mars! The rover team was overflowing the engineering analysis area, staring at the monitor in the corner. Cheers came from several spots around the floor, everywhere there was a TV monitor with a cluster of people around it. The pictures were like looking through a peephole at a new world. Some of the pictures were highly compressed, so they were particularly blocky and difficult to interpret. We saw sand, rocks, parts of the lander. The view from the IMP camera was so narrow that it was hard to know what we were looking at. Before you could figure one image out, the next one would take its place. I saw a piece of the lander that I didn't recognize, and felt a surge of anxiety while I wondered if it were a damaged or bent component. I'd never paid attention to raw IMP images during any of the ORTs; it wasn't a surprise that some images of the lander would appear strange to me. There flashed a view of sand with a rock behind it. I said something like "There's a place we can drive to!" Brian Wilcox wasn't so sure. And I realized that I had no idea what scale that rock was: Was it big and far away or small and close?

There was nothing recognizable to provide context. When the imaging engineers had assembled all of these "postage stamps" into a mosaicked panorama, then we'd know what we were seeing.

Some of those images went together to form the airbag assessment panorama. Just as in one of the ORTs, there was airbag material on the rover petal. Here was a problem the Pathfinder operations team had practiced solving. Someone on the team pulled up and modified the old ORT lander sequence to lift the rover petal up about forty degrees, run the airbag retraction motor, and lower the petal back down again. The commands were sent. We would see during the next communications pass whether they had done the job.

We hadn't gotten quite as much data from the rover as we had expected. Matt Wallace was rushing off somewhere and wondered if I had any ideas why we had only gotten data from a few of the commands in the rover sequence. I assumed it was just that the communications window to Pathfinder had closed down before the other data came down, or that the rover telemetry had been missed because it had been sent down before the DSN had locked onto the signal. I didn't see how to find out more until the next downlink opportunity.

The next communications pass showed us that something was definitely wrong. For some reason, the rover and lander were not talking to each other, or at least not well. As the existence of the communications difficulty spread through the team, all of our fears began to surface. The telecomm subsystem was the area most of us had had the greatest reservations about. The only thing we knew for sure at the moment was that we would not be deploying the rover on sol 1.

The Pathfinder folks were concerned too. So far, sol 1 had been one success after another, almost as if someone had expertly choreographed it. Even the draped airbag problem had been neatly solved by the recovery sequence, as proven by the latest IMP images. The rover communications problem was the first sour note. Jake Matijevic was forced to go off to the Pathfinder press conference at the worst possible time: We knew we had a problem; it could mean the end of the Sojourner mission if we couldn't fix it, and the rover team hadn't even had time to hypothesize on the cause. With his chronic asthma, Matijevic had never been high-energy, but tonight as he headed down to von Karman Auditorium, he actually shuffled, stooped over and feeling the weight of the complete failure that his system engineer's mind could not help but anticipate.

The rover team stayed behind to work the problem. Lin Sukamto, now Lin van Nieuwstadt, had flown in from the Netherlands to support operations. Since she had been absent during the past six months of training, the most she could do was go around and ask questions about what was going on. Now, with the radio modems she had once been responsible for in question, she was suddenly in a position where she needed to be providing answers. We all crowded into the engineering analysis area. Even Donna Shirley had joined us. The downlink team gave their report. Everybody was offering their own theories as to what was causing the problem. Van Nieuwstadt thought that the oscillator crystal had fractured on landing. Someone else wondered if the rover had gone into the silent state we used when doing APXS data collections. As the theories flowed, I added my two cents: "Everybody is tossing out ideas about what's wrong. Let's start writing them down, categorizing them, and evaluating them. The problem could be caused by hardware, software, or even an operations procedure. Let's get all the possibilities written down and determine how we're going to investigate each one. There's no point in spending a lot of time on the hardware failures, since there's nothing we can do to fix them. And it may be crazy, but is there any way that the problem could have nothing to do with the rover? Could it be on the lander? Is there some way that the lander could be incorrectly processing our data, or that it's got the rover packets, but it just isn't sending them down?"

The late-night pizza arrived. As I thought through what might be going on on Mars, I realized that my appetite had returned. I felt hungrier than I had in weeks. I grabbed my share of pizza. It occurred to me that I should be depressed about the impending doom of Sojourner, the end of everything I had devoted so much time to for the previous four years. Instead, I was happy. The waiting was over. Here was a real problem to solve, not the hundreds of imaginary ones that had been plaguing me for months. Solving problems, or helping others to solve them, was what I was trained for. This I knew how to deal with. The pizza was good.

✳

Jan Tarsala wasn't assigned to work his first shift on the rover telecommunications team until sol 2. He came into JPL anyway to be where the action was. He stayed around to see the first images arrive from Mars. In the late afternoon he went home to his family and had dinner.

About eight-thirty in the evening the telephone rang at his house. He went out to the kitchen to answer it. He knew it would be either Lin van Nieuwstadt or Scot Stride. No one else would be calling him on the Fourth of July. He picked up the phone. It was Lin on the other end, "and she was crying. She was extremely distraught." Van Nieuwstadt said, "The radio. It's not working."

"What do you mean it's not working?"

"We're hardly getting any packet throughput. We transmit and we transmit and we transmit, and nothing's coming across the data link." Van Nieuwstadt was just beside herself. "What are we going to do?"

Tarsala remembered well the predictions of Jim Parkyn, nearly a year before. "Hold on. Hold on. It's just off-frequency. We can manage that. It's not a big deal. It's going to be okay." Van Nieuwstadt feared a catastrophic failure of the radio. Tarsala tried to reassure her that if something was getting through, then the radio itself was fine. "If it's off-frequency, we can push those crystals around in frequency by heating, or not heating, the radios. We've got a heater on each end of the link. We have that control system available to us." He was convinced that they had the means to control the crystal frequencies to get the lander and rover matched up and talking again. "We can make this work."

Van Nieuwstadt had been totally uninvolved in the rover since her departure from JPL about nine months before. Yet she had so much of herself personally invested in the Sojourner radios. Van Nieuwstadt was in the worst of situations: She felt fully responsible for rover communications, but she was facing the crisis cold, without the benefit of careful preparation over the prior several months.

The rover team gathered for its midnight brainstorming session. Donna Shirley offered to be the recording secretary, writing down ideas on the blackboard. After an hour of discussion, the board was filled with possible scenarios, and annotations filled the spaces between the lines. Initially on sol 1 (and before that in cruise) the rover telecommunications subsystem had been performing just fine. Then communications degraded until almost nothing was getting through, except for garbled communications frames. We needed to be careful not to make unwarranted assumptions: Just because the symptom of the problem was no data coming across the link, this did not guarantee that the culprit was the telecommunications subsystem itself. Perhaps something had gone wrong elsewhere on the rover, preventing the rover from communicating properly.

Early in the meeting, all of the old problems had resurfaced as possible causes. These were the ones that the rover team had identified during the design, development, and test phases and had in most cases characterized, corrected, or built work-arounds to deal with. What if we hadn't done a good enough job? Someone mentioned the rover CPU crystal problem, which had once caused the Marie Curie rover to run in "slow motion." Henry Stone insisted that this was highly unlikely. The electronics team had redesigned the clock circuit. And besides, even when the problem had existed, it only occurred when the rover started up. "The rover never started out okay and then went bad!" And that was just what had appeared to happen during sol 1.

Could a temperature difference between the radio modems in the lander and the rover have caused the two radios to be transmitting at different frequencies? This was the same problem that the temperature-compensated crystal oscillators would have solved, if they had been installed. The telecommunications team had examined the issue ad nauseum over the prior two years. Lin van Nieuwstadt didn't see this as a likely cause of the communications failure. Jan Tarsala had not been told of the brainstorming session, and the rest of us had no inkling of Tarsala's confidence that temperature-induced frequency drift in the radios was indeed the root of the problem. The impression in the room that night was that while this might have contributed to poor communications, it didn't seem likely to account for the almost total loss of data transfer we were seeing.

The rover didn't have enough power to operate both its radio and the APXS at the same time, so we had designed a "silent mode" that turned off communications during APXS data collection. Most of us had been a bit skittish about the rover's silent mode ever since we'd come up with it. Any time you sent a command to tell the rover to stop listening for new commands, you had to be careful. I knew that every sequence we sent to the rover was supposed to turn the silent mode off before it completed, whether or not we'd ever turned it on in the first place. But what if the communications-restoring command had somehow been skipped? And if the rover was somehow in silent mode, where were the garbled communications frames coming from? The rover wouldn't be talking at all, so any data received by the lander would have to be spurious radio noise coming from some other source on the lander. You might expect a few random bad frames, but nowhere near the number that the lander's diagnostic software had recorded. The communications team didn't buy this theory.

What if there had been a hardware component failure? Could the lander-mounted modem power supply have gone bad? That didn't seem too likely, since there were a few frames getting through. The power supply was one of the few areas where there was a redundant component, since there were two power supplies, one to operate the modem alone, one together with the modem heater. If we ran out of other options, we could always switch over to the unused power supply and see if the problem went away.

The severity of the problem had grown worse as sol 1 progressed. What had changed on the lander during that period? The obvious thing was that the lander's high-gain antenna had deployed and was tracking the Earth all day long, maintaining communications between Pathfinder and the DSN. Perhaps as the high-gain antenna moved to new orientations, it began to scatter or block radio signals between the lander-mounted rover antenna and Sojourner, getting more and more in the way as time went by. Since the rover was still stowed on the lander petal, its antenna was also still lying down along the solar array, while the lander-mounted antenna was deployed and vertical, leaving the two antennas perpendicular to each other. Although data had gone back and forth between the rover and lander during ground testing, this was the worst possible relative orientation of the antennas. If other conditions had weakened the link, then this might have pushed the telecommunications subsystem over the edge. It also meant that once we managed to get the rover unstowed and its antenna upright, the current problem might go away.

Other than one or two suggestions and questions, there was little for me to do at the meeting but listen and look for holes in the logic. The subsystem experts were far better than I at inventing hypotheses to explain these symptoms. I had done my job: nudging people in the right direction, then standing back to let them do what they did best. Yet I was proud to be sitting in that meeting. For most of my life, I had been amazed by the stories of JPL engineers who could magically figure out what was wrong with spacecraft far out in space and, even more impressive, could fix them! Here I was, and I was making my own small contribution to just such a group of engineers. I didn't know if we would be able to fix rover communications. I didn't know if it could be fixed. But if it was possible, the people on this team were the only ones to do it. Most of my thoughts were focused on the discussion. But a small part of my mind would intrude periodically, announcing silently, "Damn. I love my job!"

With all of the options listed, the next steps would be to investigate each one and determine which could be discounted, leaving the rest as possibilities. Lin van Nieuwstadt could barely keep her eyes open. She was still on Netherlands time, so she should have been in bed twelve hours earlier. Around 2 A.M. Matijevic ordered himself and the rover team to go home. There wasn't much more they could do tonight, and he felt the team would think better after something like a good night's sleep.

After the exhausting day and night of sol 1, Matijevic had hoped to stay home most of the next day and get some rest. He woke up mid-morning on July 5 and turned on the TV. Since there was no other bad news to report on Pathfinder, the only topic seemed to be the silent rover. There were speculations that the commercial radios were no good. Matijevic sighed and shook his head. He'd better get into work and shield the rover team from the media so they'd actually get the time to solve this problem . . .

When the first downlink for sol 2 came in, there was lots of rover telemetry in it. A total of 31,491 bytes had been transmitted, which happened to be the capacity of Sojourner's telemetry buffer. Apparently, during the period of failed communications on sol 1, the rover had continued to execute its command sequence as ordered, dutifully storing away its telemetry until it overflowed the buffer memory. We'd lost whatever data the rover had generated after the overflow, but at least we were talking again.

When the press heard that the rover was working, they wanted an explanation. The real answer was that we didn't yet know. But between the media's need for an answer and the attempts to accommodate by very pressured engineers, pure speculations were transmuted into apparent facts. The story went that the rover had been fixed by the equivalent of "hitting the reset button on your computer a few times." This version of the truth spread so far and so fast that even spokespeople at JPL, as well as members of the Pathfinder team, believed it and were repeating it to the press. I knew that we hadn't yet had the opportunity to do anything unusual to rectify the problem. The only "resetting" of the rover was due to the normal command sequence already onboard, which was designed to have the rover shut itself down overnight, waking up once per hour to get temperature readings. I doubted that that had fixed anything.

Scot Stride winced when he heard the press reports. The rover telecomm subsystem was his, Jan Tarsala's, Sami Asmar's, and Lin van Nieuwstadt's, and as far as he was concerned, they were still in the process of assessing the problem. Stride's own guesses as to the cause stemmed from the knowledge that they had never had the chance to test rover-lander communications in the particular lander configuration that now existed on Mars: The rover and its antenna were stowed with the lander petals open, sitting in the middle of a wide open terrain. The closest they had come to this on Earth had been during the system thermal/vacuum test, but at that time the lander was enclosed by metal chamber walls, which caused the radio signal to bounce all over the place.

Whatever the cause of the communications loss, and whatever the reason that communications had been regained, it was now time to deploy the rover.

*

Justin Maki was a research associate at the University of Arizona. Officially, that put him on the faculty, but in reality he was barely one step up from graduate student slave labor. Peter Smith had hired him to build IMP camera imaging sequences during the mission, and brought him out to JPL with the rest of the IMP team to do the job. Here he was, barely twenty-seven, just out of school, and he was part of the Pathfinder mission! He couldn't believe his luck. He had a VOCA unit on his desk, so he could hear everything that was going on in the Mission Support Area, and his office was next door to Rover Control. Pathfinder was on Mars, the IMP camera was working like a champ, and his sequences were telling it what to do. It didn't get much better than this.

Well, there was one thing he did want to do—make rover movies. Maki figured that if he knew how fast the rover moved, and he could get the timing close to right, he could track the rover with the IMP, and take pictures a few times a minute while the rover was driving. If you put photos together, you'd get a kind of jerky movie showing what the rover did. The resolution of the images wouldn't have to be so good, either, so the data volume hit would be small. The problem was, the Experiment Operations team might see such rover movies as superfluous, having no scientific value. But Maki knew that rover movies would be a great way to document what Sojourner was doing on Mars, and would probably capture the imagination of the public. More important than that, it would be fun to see if it could be done.

Before landing, when the last few Operations Readiness Tests were going on, Maki had approached Bill Dias, the project's primary Surface Planner. Dias was responsible for designing the scenario of how the mission would unfold after landing. He was also responsible for budgeting the downlink data volume between the competing sources of telemetry, which included the lander engineering subsystems, the IMP camera, the lander weather experiment, and the rover. Dias liked the idea of rover movies. In fact, he had always maintained an extra margin of telemetry volume in case more imaging of the rover proved necessary.

Dias and Maki schemed. What was the best way to incorporate rover movies into the plan without being too obvious about it, which might get the idea quashed? A rover movie would have to be a command sequence of its own. Every sequence needed a unique identifying number. For example, the imaging of the rover taken every day to update the rover's position was always in sequence S-0055. The status of all sequences was reviewed at the Mission Plan Approval meetings scheduled for every sol. Anything too different from the other sequence numbers might stand out, calling potentially fatal attention. Maki and Dias went through the sequence assignment list. The 50 series was mostly rover-related sequences. S-0053 and S-0054 were reserved for soil mechanics experiment imaging, S-0055 was taken, and S-0056 was for imaging the tracks the rover would leave in the Martian soil. S-0050 was available. Maki smiled. "Let's call it 'Rover Navigation Imaging.'" That sounded a lot more respectable than "Rover Movie."

During ORTs 6 and 7, Maki had experimented with movies of the rover egress down the ramps. Getting the timing right seemed to be pretty easy, but aiming the IMP was a bit of a problem. The first time he tried it, he caught only the top half of the rover, then only the wheels. And aiming the engineering model of the IMP camera was a bit different from aiming the camera on the actual spacecraft. He would just have to give it his best shot when the real opportunity came along.

✳

We had uploaded the sequence that commanded the rover to stand up. I was looking at the monitor in the engineering analysis area when the telemetry started coming in. The IMP image of the rocker latch flashed up on the screen, and was gone, replaced by the next picture. But the latch spring had been straight and true. I couldn't vouch for the right rocker, which could not be seen with the lander camera, but the left side had deployed just fine. My worst nightmare looked like it would remain only that.

But then the engineering telemetry from the rover itself indicated an error. Henry Stone was the Data Controller on watch, and he would later remember this as one of his personal defining moments for the mission. It was his job to decide whether the rover stand-up had been successful. "It's your call, Henry," Art Thompson said to him over the VOCA net. If we were to get the rover off the lander during sol 2, we'd have to make the decision soon. But Stone didn't have enough data yet to be sure! He was keenly aware that if he made the wrong choice, he could be the goat of the mission. If the rover backed out of its wheel restraints with an un-latched rocker, there was no going back: There would be no way to lock that rocker in place. The rover mission would be over then and there. From the IMP photograph of the rocker-bogie, Henry knew that the left rocker had latched properly. Digging into the data, it became clear that the error was being reported for the side of the rover that we could not see, and indicated that the encoder on the bogie had shifted a few counts compared to what we had expected. But that might be explained by the fact that this was the first time the FUR had stood up in Mars gravity. We were so cautious in our test for success that only a slight variation in the expected readings would cause the rover to stop what it was doing and wait for human intervention. "Okay, Henry. Don't be rash," he told him-self. Howard Eisen was already convinced that the rover had unstowed successfully, and was vocal about that opinion. While Stone was still not happy with Eisen's certainty in the perfection of the unstow, he generally agreed that, if the unseen rocker-bogie had not locked, the backup latch test would have caused that side to sag noticeably, giving the rover a skewed appearance in the images and unmistakable discrepancies in the potentiometer readings. None of that had happened. Stone declared the rover healthy and ready for egress to the surface.

Earlier, Brian Cooper had been studying the stereo IMP images of the ramps in the RCW display, and had come to the realization that the for-ward lander ramp was sticking out into the Martian air, suspended like a diving board over the ground. We were going to have to drive the rover down the other ramp. Going down the rear ramp looked like a good choice. It led to a clear, flat, open area almost completely devoid of rocks. The one rock in the vicinity, which had been immediately christened "Bar-nacle Bill" for its mottled appearance, would be a perfect first rock target for the rover's APXS. Rick Welch built the modified rear-egress sequence. Justin Maki designed the first real "rover movie" sequence for the IMP, intended to capture a special moment on Mars, with the timing set to correspond to the delays inherent in the rover command sequence. Maki had no sophisticated software tools to help him select the appropriate pointing parameters for the IMP, which depended on how steep the rear ramp was tilted. Instead, he eyeballed it, counting the number of rows of pixels in the existing ramp pictures to guess how much further he needed to tilt the IMP to point it right at the end of the ramp. The sequences were uplinked to the lander.

When the thumbs-up came from the downlink team, the authorization went to the Coordinator and on to the Flight Director. The commands went up to queue the sequences already onboard the lander. The team waited for the confirming telemetry.

The rover movie images started to come down, mixed together with the other IMP images. Every one in the MSA watched the screen. Maki waited intently. Had he timed the imaging correctly? Had he aimed the IMP at the right spot? The first image told him that he had. The end of the ramp was in the center of the view. The rover would have to come this way to reach the surface. But there was no rover in the picture. That wasn't too surprising: Maki had designed the sequence to have two images before and after the rover should have driven through the scene, just in case the timing was off. A few IMP photographs of other targets appeared. Then the next ramp image flashed up, still with no rover. Okay. Next one. Still no rover. Maki worried again. Had he missed it? The rover was late. He did not want to have messed up. Another member of the IMP team from the University of Arizona pointed a finger at Maki. "You screwed up. You screwed up!" he repeated, an odd smile on his face.

Meanwhile, Brian Cooper looked at the same pictures and panicked. Shit! Where was the rover? In his mind, the worst case scenario played out: The rover had rolled off the side of the ramp, flipping over to lie permanently upside down in the Martian dust. Where was it?

At the same moment, Henry Stone stared intently at his workstation screen in the analysis area. He was waiting for the rover engineering telemetry that would either dispel or confirm his lingering doubts over the success of Sojourner's stand-up. What if the right rocker really hadn't latched? Then they'd be screwed. Some telemetry came in. The first "MOVE" command, driving Sojourner several inches onto the rear ramp, had executed normally. There was no error report. Stone relaxed: Both rocker-bogies must be fully latched, or the command would have failed. He had made the correct choice.

Art Thompson, the Rover Coordinator on duty in the MSA, was also watching the engineering telemetry stream, waiting for it to confirm that all of the egress commands had executed without error. The Flight Director was impatient. "Rover. Can you confirm?" The rover telemetry stream had stopped at command 2679. That only told Thompson that the rover had backed halfway down the ramp without detecting a problem. There was no more telemetry! The damned communication problem was still there. Thompson pondered what he was going to say. He really, really didn't want to have to say he didn't know what the rover had done. He paused. Someone told him to look up at the monitor displaying the IMP images as they came in.

There it was! The rear left wheel appeared in the next image. The rover was nearly at the bottom. Then the wheels were on the ground. Finally, an image flashed up with Sojourner fully off the ramp and safely on the surface. The MSA was filled with cheers. All right! Thompson spoke into his headset microphone. "We can report visually six wheels on soil!"

The rover had indeed been about a minute late: As a consequence of the rover-lander communications problem the rover had spent more time than usual attempting to communicate before it drove down the ramp. But now it sat firmly on the ground, ready to carry out its mission. The first thing it did was plop the APXS down onto the soil and begin collecting data.

Maki pieced together the ramp images that comprised the rover movie, and made it available on the Pathfinder network. He went next door to Rover Control and told us to take a look at it. We called up the rover movie and played it on one of the RCW displays. It was just a few frames, repeated over and over again, but we didn't tire of watching the rover drive down the ramp onto the surface of Mars. The movie revealed one thing that nobody caught when the images were displayed one by one during the downlink: In the frames before the rover came into view, you could see the ramp flex under the weight of the advancing rover. As we were attempting to glean all we could from the few images that we had, someone from the MSA rushed in and told us to get out of the way, that we were blocking the view. JPL was sending the rover movie out to the public, and the way it was being done was this: There were three live video feeds in the Pathfinder Mission Control area—one in the MSA, one in the big conference room, and one in Rover Control. The rover team had dubbed its video camera, mounted in the back of the room on a remotely operated pan/tilt platform, "HAL9000" because its big zoom lens was reminiscent of the computer's "eye" in the movie *2001: A Space Odyssey*. The camera operator, wherever he was, had noticed the rover movie on the RCW display, and had zoomed in for a close look. Of the three available feeds, this was currently the most interesting, so it was being fed out to NASA TV, and from there to the public at large. So, unbeknownst to those of us in Rover Control, the public was looking over our shoulders at the same thing we were. Or at least they were trying to. Apparently, Sharon Laubach was standing in front of the SGI, partly obscuring HAL9000's view of the movie. And, thus, someone was dispatched to keep us from interfering with the show!

Laubach found out later that her brother had been watching. He was in the U.S. Navy, stationed in Japan. He and his buddies were watching the NASA feed, and he had told them that his sister was working on Pathfinder. Every time a woman appeared on the TV screen, someone would ask, "Is that your sister?" and he would respond "No, that's not her." Finally, somebody said, "Is that your sister's nose blocking our view of the rover?" "That's her all right!" At least her nose was famous, seen around the world.

*

At the sol 3 uplink planning meeting, Jan Tarsala explained the radio frequency-drift hypothesis, and how it could fully account for the on-again, off-again communications performance we had observed. The solution would be to properly control the temperature of the rover's radio.

The big question was whether to run the heater to warm up the rover's radio, or let the radio stay cool by leaving the heater off. Tarsala didn't yet have enough telemetry from the rover to be sure which was the right choice. His gut feeling told him to keep the radio cool. He made his recommendation to Henry Stone, Art Thompson, and me: "Don't heat the modem on the morning of sol 3."

I held up a sheaf of papers. "Here's the sequence. It's already built. It's up there right now, running on the rover. I've already got the heaters on. When the rover wakes up tomorrow morning, it will heat the radio for ten minutes." There was nothing I could do. There was no way to comply with Tarsala's request. The heating of the modem early on sol 3 was con-trolled by commands near the end of the sol 2 command sequence, before the rover ever asked the lander for its new sequence. It had to be that way, because the rover needed to communicate to receive the sol 3 sequence in the first place.

So turning on the heater on the morning of sol 3 would be a kind of experiment. Tarsala's bet was that the communications link would get worse as the rover's modem got warmer: "Fortunately, I was dead wrong," he said later. Sol 3 communications looked good. As the next few sols would show, the radio needed heating, not cooling. "Once the radio exceeded a certain temperature, things got much better very quickly."

Within a few sols, rover sequences would contain commands to pre-heat the radio before critical transmissions. Whenever possible, image transmissions, which would have long transmit periods, were moved to later in the day, when the radio would naturally be warmer. Rover com-munications turned manageable.

※

The post–sol 2 press conference was triumphant. The entire rover team had been invited to be present in von Karman Auditorium. On Earth, at JPL, it was late in the evening of July 5, 1997. Warm though it was, I was not going to deny myself the opportunity to wear my leather rover team jacket. Today we had earned the right to pat ourselves on the back, if only for a few moments. As we gathered in the auditorium, everybody on the rover team was shaking hands. Donna Shirley came over and hugged sev-eral of us. She made me turn around to show off the unofficial rover team patch on the back. I gave the jacket over to Jake Matijevic as he sat down on stage, in case he needed a prop during his presentation. Sharon Laubach arrived with an 8×10 of the first picture taken by one of the rover cameras; it had just come off a photo-printer near the MSA. She handed it to Matijevic and then joined the rest of us who had already congregated by the Pathfinder lander mock-up on the west side of the room. This was the rover's night.

I had sometimes been concerned that Matijevic was not the most dynamic of speakers. He tended to speak in complex sentences full of clauses. His condition caused him to cough often, and sapped his energy. Tonight, I need not have worried. Matijevic was more excited than I had ever observed. He spoke of the new planet to be explored and the fully functional rover that was going to do that exploring. He surprised me further by actually holding up the jacket and speaking about the people who had made the rover a success, and then asked the audience to acknowledge the rover team. The rover engineers stood to one side of the auditorium interspersed among many members of the Pathfinder team. Matijevic's commentary had shifted the attention to us: As the audience applauded, the video cameras turned in our direction, scanning the faces of the team.

✳

Where was Bill Layman during all of this? By the time of the Pathfinder landing, Layman had been off working on another project for nearly a year. He had not been involved with operations training, but had been providing his special expertise elsewhere, where it was most needed. On July 4, he and his wife watched the landing and what followed on TV at home. He had told Howard Eisen and others that he would be sitting by the phone, ready to help out in any way he could, if there was a problem. Layman felt that he hadn't paid his dues during that year of preparation for surface operations. "I am struck by how many politicians come out of the woodwork to touch the heroes. It's like the game-winning pass—everybody wants to touch the quarterback. I had that same feeling, that it would be fun to bask in the glory. But it was really only the guys who had trained throughout that final year who ought to be there. Other people, particularly me, shouldn't be standing around, in the way, mucking things up." He was also keenly aware that for the bulk of the Sojourner design phase his decisions had been the definitive word on the rover, and had determined the direction in which the team marched. He was afraid that if he were on the floor, in the middle of things, people might turn to him for answers and accept his responses as gospel, when in fact his knowledge was a year out of date. He felt he was now less educated on the key issues than the well-trained members of the rover operations team. He stayed away.

Once more, Bill Layman had made a decision with only the good of the team and the mission in mind, demonstrating the strength of his leadership in a way most would never recognize.

Yet, though Layman would not allow himself to be present in the center of mission operations, the success of Sojourner was, by his own reckoning, the high point of his technical career. He had been the project mechanical engineer on both Voyager and Galileo, JPL's flagship missions. But those spacecraft, flying past and orbiting distant planets, just could not match the sheer emotion triggered by Pathfinder and Sojourner on Mars. Time on the Pathfinder project was compressed, the events following so closely upon each other: landing, the first signal from the surface, the first images of Ares Vallis, the rover driving on the alien terrain. Pathfinder delivered huge, instant gratification. The microrover team had done it! The team he had prodded, challenged, encouraged, and finally been forced to leave behind as he went to help develop the lander that would deliver Sojourner safely to the surface; they had together created this. Layman had always felt a great responsibility to the team, and had never been sure that with the conflicting demands of the rover and lander on his time that he had totally fulfilled that responsibility. The thrill of accomplishment and release of years of effort were overwhelming. There were tears in his eyes. Even a year later, recalling the images of the rover rolling onto the surface for the first time, the tears nearly came back. "What a rush. We did it. And nothing can ever take that away from us. What a thrill to have been the Chief Engineer on the rover! That was really something!"

TWENTY-FOUR

LIVING ON MARS TIME

Sol 3

"Something's screwy!" Henry Stone and Matt Wallace were not happy. I had just walked into the engineering analysis area to find out what the rover's telemetry showed for sol 3. We were only just beginning to get a handle on the communications link mystery. Now something else was obviously wrong, and they did not yet know what had caused the problem. For the night of sol 2, we had commanded Sojourner to collect an APXS spectrum of the soil behind the rover while it slept. The plan was to wake up the rover every few hours, store the intermediate APXS results in the rover's memory for later transmission to the lander, do a health-check, and shut down again. This would satisfy the APXS scientists that their first APXS spectrum of an actual piece of Mars would not be ruined by some "glitch" that introduced noise into the data. The healthchecks would also give the rover team a set of temperature measurements, spaced out over the night; from those data points, we would be able to re-fine our thermal model, and come up with a heating strategy to make sure the components in the WEB stayed within a survivable temperature range.

But almost none of the planned night operations had occurred. In the late afternoon, Sojourner had shut down, then collected APXS data for three hours in night mode, and powered on again. The rover was supposed to shut down again for a couple of hours, then power on to take another reading. Instead, at 7:08 P.M. MLST, Sojourner went to sleep for the rest of the night. When the rover finally woke to the morning light, the scheduled times for all of the nighttime shutdowns had passed, so Sojourner skipped all of those shutdowns. The temperature readings, intended to be spread across the night, were all bunched together in the early morning.

Stone was speculating that maybe there was a hardware problem with the rover's clock. He just didn't see an operational reason for the rover to shut down to an apparently random time. Once I had gotten them to explain the symptoms to me, an obvious possibility came to mind. "What time was it when the rover shut itself down? Was it after the rover's internally set 'END-OF-DAY' time? I think it may just have done an auto-shutdown." From the expression on Matt Wallace's face, I knew the thought had never occurred to him. Meanwhile, one of the downlink engineers had gotten Jack Morrison on the phone. He zeroed in on "auto-shutdown" as the culprit even faster than I had. The auto-shutdown had effectively inserted a "sleep until 8 A.M." command in front of the other shutdown commands in the sequence.

Now the question that remained was why had we never seen this before? Why hadn't auto-shutdowns interrupted our sequences during ORTs? We had done many overnight wakeups during testing. Morrison had the answer for this too. When the conditions were right for auto-shutdown (when solar power was low and it was past Sojourner's bedtime), it would still take about two minutes to trigger. That was normally more than enough time for the rover to wake up, readout APXS data, perform a healthcheck, and shut itself down normally by command. The problem on the night of sol 2 could be traced straight back to the low performance of the rover-lander communications link we had been seeing. With communications still spotty, the rover had built up a big backlog of telemetry data in its buffer. When Sojourner woke from its afternoon nap, the first thing it did was send that data back to the lander. This used up the two minutes, and auto-shutdown took over before the commanded shutdown could execute. We could avoid the problem in the future by turning off auto-shutdown whenever we chose.

Later that day, Wallace acknowledged my help in identifying the problem. "That was a good call." He probably thought I really knew my stuff. What I knew was that I'd scared myself almost a year before when I thought a mistimed auto-shutdown in the middle of the day might ruin sol 1 . . . I couldn't help but think of it. Sometimes your past mistakes don't come back to haunt you, but to help you.

✳

Even though we hadn't gotten all the intermediate results the APXS team had asked for, the one spectrum we did get was a clean uninterrupted fifteen-hour integration. We'd bagged our first APXS spectrum of soil!

And even as the engineering analysis team was struggling with the mysteries of auto-shutdown, we had uploaded the sol 3 command sequence to snag Barnacle Bill. In the planning meeting, the Experiment Operations Working Group had been unanimous: Get Barnacle Bill! It was the obvious choice. There were no other rocks for yards, Barnacle Bill was almost the right size, and its mottled appearance made it an intriguing target. When Brian Cooper had surveyed the end-of-day pictures for sol 2 through his 3D goggles, he was impressed with his good fortune. The images showed the rover's position at the bottom of the ramp, just after egress. Barnacle Bill was already so close and well aligned with Sojourner that Cooper would not even need to back the rover farther away from the lander; all Sojourner had to do was turn seventy degrees to its left, partially deploy the APXS to "ramming" position, and drive backwards just about one foot. He checked the rover motions again and again, but they looked right. I merged the traverse commands into a sequence that also did a soil mechanics experiment in the morning (digging the front left wheel into the sand) and some imaging.

When the afternoon results from sol 3 came down, we realized we had just had a great day. Sojourner had hit Barnacle Bill on the first try. Lots easier than our ORTs! The performance of the radio modems was the best since landing. And the rover sent down its first picture of the lander seen from off-board. The image showed mostly airbags, but it was beautiful nonetheless.

✳

At one of the press conferences a reporter asked, "There are those who have looked at pictures and say 'Gee whiz, this looks like Arizona . . .'" Have we really landed on Mars, or is this out in the desert somewhere? Matt Golombek just thought the question was silly. "Here's a place that does not exist anywhere on the Earth. It's as dry as Arizona and it's as cold as the arctic. There's nowhere on the Earth with this temperature, pressure, and these sorts of surface features." Tim Schofield was the lead for the weather experiment on the lander. His response was that he "certainly wouldn't be staying up all night to watch fake data coming in!"

The landing site seemed too perfect to have been faked. No secret group of scheming engineers would have designed such an ideal location, and if they had, it would have cost more than going to Mars in the first place. The end of the rover ramp let out into a large flat region nearly devoid of hazards, just the right spot to get our "Mars legs" and get used to driving on Mars. Beyond the open area, when the rover was ready, was more challenging terrain with a number of rocks to study, including one set grouped closely together that had quickly been dubbed "The Rock Garden." Even better, the site had given us Barnacle Bill—the perfect APXS target rock, right where we wanted it. What more could the rover team ask for?

✳

Over the first week, the Data Controllers learned to give the uplink team what it needed. At first, Allen Sirota and Henry Stone seemed more focused on getting the rover status reports complete and on the web than in sifting out those details that were most important to operations. The reports were loaded with data. The Data Controllers would complain about how difficult it was to get everything done in the limited time available. Those of us on the uplink side got more and more frustrated. We didn't care about all the details! We just wanted the critical few facts that would tell us whether we could operate the rover normally, or if we had to do something special, like heat the radio modem or tell the rover to go to sleep early to prevent overheating. As the Mars days went by, we settled into a pattern, and the transfer of data between team members improved.

Sol 5

At the end of sol 5, Sojourner completed its primary mission objectives, two days early. In prior sols, Sojourner had performed a full set of technology experiments, and collected both APXS soil measurements and Barnacle Bill rock measurements. On this sol, Sojourner delivered a beautiful image of the Pathfinder lander. The entire lander was visible, except for the IMP camera, which rose up out of the field of view. The image revealed Sojourner's own path, as evidenced by wheel tracks leading in an apparent straight line from the extreme foreground back through the mottled soil to the deployed ramp.

The rover team declared success. We were now in the realm of the rover extended mission. (We could thank Donna Shirley for negotiating a practical set of mission success requirements years before.) But other than being a nice thing to contemplate in a spare moment, no one noticed. We went back to work.

Sol 6

At the sol 6 press conference, Peter Smith began his presentation of the latest IMP images with a typically dramatic introduction, a poem which he called "a quick observation on the effects of shifting your schedule thirty-nine minutes later every day to stay on Martian Local Solar Time":

When you say good morning as the sun is setting, now that's living on
    Martian time.
Your sunglasses look like this [Smith put on a pair of red/blue 3D
    glasses.]—that's living on Martian time.

No time for laundry and you get your shirts out of the box of project
    T-shirts every day—that's living on Martian solar time.
When you start admiring strange-looking rocks, giving them names and
    talking about them to your friends—that's living on Martian time.
When your days are called sols and your nights are called days—that's
    living on Martian time.
But when you start laughing at the engineers' jokes—you know you're
    living on Martian time.

The audience loved it. And Smith had identified a fundamental truth:
The mission operations teams were living a skewed, warped, intense existence. They had lost all sense of the normal flow of night and day. In the
center of all of the media attention, we were nearly completely isolated
from the rest of the world.

I would get phone messages from friends and family, congratulations
on the success of the mission. But I could never respond, because the few
lulls in the rush of activity would inevitably occur in the early hours of the
morning.

Eating was a problem. My appetite was back, but I was never hungry
at the right time. On sol 1, I had come on shift about 10 P.M., Pacific Daylight Time. Lunchtime started out around 2 A.M., and got progressively
later. Even the pizza places weren't open then. And since mealtimes were
changing every day, I was never quite sure when my body would be ready
to eat.

Sleeping was a problem. I might finally finish reviewing a sequence
and head out of the Space Flight Operations Center at 10 A.M., on a bright
sunny warm day I had no time for. I'd drive home, exhausted yet keyed up.
I would try to sleep in the middle of the day. I would succeed for a few
hours. Then I'd wake up wondering what was happening at work. On one
of those days, I went downstairs, turned on the TV, and surfed through
the channels. There was one education channel running the live feed from
JPL, all the time. When I found it, I stopped and listened to the voices in
the MSA. I heard Matt Wallace's voice. "Flight. Rover." He was asking for
the attention of the Flight Director. There was no response to him.
"Flight. Rover," he said again. The Flight Director was listening to some-

body else. "Flight. Rover." As usual, Wallace had more patience than I did. "Answer him already!" I yelled at the TV. "I want to know what's going on!" Eventually the Flight Director responded, and I could relax, knowing from the dialog that the rover was still fine.

It was a fishbowl existence. When else in my life would I be able to turn on a television set to watch the minute-by-minute goings on at work? Somehow, the project I had worked on for so long had merged momentarily with the national consciousness.

A few days after landing, a story began circulating around the building. Sometime around the second day of surface operations, an urgent telephone call for Wes Huntress had come in to JPL. Huntress was the NASA Associate Administrator for Space Science, head of the office within NASA that had funded Pathfinder. He was visiting JPL to observe the culmination of the mission. The call came from an important personage at Johnson Space Center, the NASA center that trained the astronauts and operated the manned program, including all of the Space Shuttle flights. Huntress was tracked down and the JSC official finally got the Associate Administrator on the phone. The official was furious that the JPL operations personnel were presenting the wrong image to the public. Where were their shirts and ties? They were wearing shorts and T-shirts, cheering and hugging, putting their feet up on counters, even eating pizza in the control room! All of this was being transmitted live to TV audiences around the world, and huge numbers of people were seeing it. Was this the impression of NASA that the organization wanted to communicate to the public? Would Huntress please see to it that the Pathfinder team shaped up and conveyed a more professional appearance and attitude? Huntress's reported response was to tell the JSC official, "Fuck off!" and hang up the phone.

Whether or not the story was true, the Pathfinder operations teams loved it. The proof of our professionalism was that Pathfinder was on Mars, the lander was functioning perfectly, and Sojourner was driving around up there! And NASA was astute enough to recognize it!

And, as the team sensed but which became clearer as time went by, the public was enthralled. People wanted to know how the rover was faring, day by day. Hits on the Pathfinder web site literally doubled the volume of Internet traffic during the early days of the mission. On the Monday after landing, when people were coming back to work for the first time since the Fourth of July holiday, there were 47 million hits on the site, as people downloaded images from Mars. Interest in the mission extended worldwide. There was a rumor that the entire French telephone/Internet network had been brought down by the volume of traffic directed at Pathfinder, and that the French government had asked its citizens to show restraint in order to restore service.

To many of the people tuning in, the young, casually dressed engineers in the control room looked a lot like themselves.

Most of the science team, involved only with lander-based instruments, took rover operations for granted. But whenever Sojourner achieved a new objective, Hank Moore would send us a congratulatory email: "Great job rover team!" The JPL store had started selling Pathfinder baseball caps with an embroidered Sojourner on one side. Once Moore got his, he never seemed to take it off. Some of the other Pathfinder scientists wondered when the rover would die, so they could stop factoring rover IMP images into their science plans. Moore would grin and reassure the rover uplink engineers: "Sojourner's a tough little girl. You'll still be driving her a year from now!"

During breaks in the action, or if we were off shift but afraid to go home because we might miss something, some of us would steal into one of the conference rooms to look at the twenty-foot-long poster of the landing site. Covering an entire wall of the room was a printout of the "Monster Pan," the full 360-degree view from the IMP camera, composited together from more than a hundred smaller images. Scientists on the Pathfinder team had hand-lettered names on scraps of Post-it notes and stuck the notes over various rocks visible in the poster. The names were a convenience: It was easier to talk about Yogi, Barnacle Bill, and Chimp, than to use a number and try to remember which rock the number had been assigned to. Matt Golombek, as Pathfinder Project Scientist, had specified rules for rock-naming, most particularly that rocks were not to be named after people. He also implied that since the names were to aid the scientists, only the scientists should do the naming. The engineers on the Pathfinder team who had any interest in naming rocks ignored the injunction. Once a Post-it note went up on the poster, no one but the namer knew the source.

There were still rocks to be named. It became a pastime with the team to search out unnamed features in the Monster Pan, and correct the oversight. Eventually the Post-its were so thick on the poster that Golombek had to assign one of the junior members of the team just to keep track of the names.

Other Pathfinder traditions spontaneously arose. Rover wakeup songs were one of these. On sol 3, Howard Eisen had the idea of playing a song for Sojourner just before the morning session during which the sol's command sequences would be uplinked to the lander and rover. Mission Control for manned space missions traditionally greeted the astronauts with a wakeup song each morning. Why not do the same for the rover? The first song played was "Final Frontier," which sounded appropriate for a deep space mission, but in reality was the love theme from the television series *Mad About You*. Over the next few sols, rover team members sometimes brought in songs to play, and sometimes forgot. But by sol 12, the habit was in place. Rover wakeup songs were being played every sol. Not having a song was no longer acceptable. On one sol, Rob Manning got into the act, playing a live solo of "Centurian Starfire" on his trumpet in the Mission Support Area. Some of the media had the impression that the songs were actually being transmitted to Mars. In reality, the rover wakeup songs were only sent across the voicenet to be heard by the rover and lander teams on the second floor of Building 230. The songs were really for them, as they began a new day operating two robotic spacecraft on another world.

The rover uplink team participated by submitting songs, but almost never actually heard a wakeup song being played. By the time of the morning uplink session, we were already at home, in bed, and asleep. By definition, the wakeup songs greeted the Martian day. The uplink team worked the Martian night.

Sol 12

On the night of sol 11, Sojourner completed an APXS experiment on the big rock Yogi. For sol 12, the science team asked us to send the rover to a whitish patch of ground they had named Scooby Doo. On the way, we would stop at a sandy area called the Cabbage Patch to do a wheel abrasion technology experiment. The experiment would cause just the right middle wheel to turn, digging it into the soil to see if the Martian sands were rough enough to scrape off the special material that had been painted onto the wheel's surface.

Up to this point, the uplink team had accomplished all of the rover's traverses using only low-level "MOVE" and "TURN" commands. We hadn't wanted to risk the autonomous "GO TO WAYPOINT" command until we'd confirmed that Sojourner's laser stripes were indeed visible to the rover cameras. On sol 8, the rover took a complete photograph of one laser. The stripe showed clearly in the image, stretching into the distance. The lasers worked. The Cabbage Patch was only a few yards away, but the rover team was unanimous that it was time to take advantage of the autonomous navigation capability we had built into Sojourner.

Peering into the RCW stereo display, Brian Cooper designated the first Martian waypoint, a modest traverse of about seven feet. Rick Welch integrated Cooper's traverse commands into the overall sequence, which included the wheel abrasion experiment and several rover images. When Welch and Cooper came in to work the next day, the engineering analysis team gave them the best news possible: nothing much to report. The "GO TO WAYPOINT" had executed without error, Sojourner was sitting in the Cabbage Patch, and the vehicle was in good health. What next?

When I first had time to stop and think, it seemed strange that rover operations on Mars were going along more smoothly than they ever had on Earth. But it was unquestionably true. Everyone on the rover team recognized it. We were getting great images from the rover cameras. We'd been hitting our targets. (The downlink team had even put together an APXS target scorecard. A single sheet of paper was posted on the wall outside the engineering analysis area. On it were the names and small pictures of each of the rocks the rover had reached, reminiscent of the tally that World War II bomber crews painted on the sides of their aircraft. The words across the top of the sheet said, "You pick 'em. We'll plant 'em.")

Perhaps the relative ease of rover Mars operations wasn't so odd. The rover team had trained for months, and the failures during that time stood out in all our minds. The engineers had learned from their mistakes. And although we had dutifully pretended during RORTs and ORTs that Marie Curie was on Mars, there was no substitute for the sure knowledge that Sojourner was now truly on another planet—far beyond rescue if we made a mistake—to add that last bit of care to everything the operations team did.

And there were still problems. Sometimes the rover halted its driving when it encountered a steep slope that wasn't really there. Sojourner's tilt sensors, intended to warn of steep slopes and impending rollover hazards, were sticking and sometimes giving erroneous readings. We didn't know why. If the Rover Driver on duty (either Brian Cooper or Jack Morrison) decided the terrain of the rover's next traverse looked safe, he could choose to turn off the offending tilt sensor.

Sojourner's turn rate sensor had adopted the unpredictable ways of its twin on the Earth-bound Marie Curie, sometimes leading the rover astray as it drove. After long traverses, Sojourner might end up a yard or more from its target. If that occurred, the rover planners consulted with the science team to decide whether to continue on to the original destination, or go for a nearby "target of opportunity." We were on Mars . . . Chances were good that some rock close to the rover would excite the scientists.

Sol 30

Around sol 30, the rover and lander uplink teams revolted. Both teams had been living on Mars time for a month, and we were exhausted. The adrenaline rush that had kept us functioning during the fourteen-hour shifts for the first two weeks had long since dissipated. The schedule hadn't hit the science teams as hard: most of the scientists only had to stick around until they had decided what they wanted the lander and rover to do during the next sol. Then the scientists could go home, or do detailed analysis of telemetry on their own schedules. Much of the science team had already evaporated. Many of them had returned to their home universities, continuing their participation by telephone and Internet.

Meanwhile, the uplink teams, still shifting their work schedules by forty minutes each day, remained largely isolated from family and friends. An unexpected consequence of a wildly successful mission—so far—was the prospect of living this Mars-centric life for the foreseeable future. Was there no way to start living on Earth time again?

The lander uplink team had one advantage over their rover counterparts: The lander didn't move. Most of the command sequences they planned were for IMP science imaging—sets of pictures taken through various filters at specified resolutions. Building a sequence a few sols ahead was feasible. If you could do that, you could mostly do the work on a regular shift.

But Sojourner did not stand still. So sequences for a given sol could not be generated without the telemetry from the previous sol. Yet the rover team was getting better and better, on both the downlink and uplink sides. Most of the downlink team had little to do, as long as none of Sojourner's hardware suddenly failed. In the absence of vehicle anomalies, the Data Controllers could handle a first-cut downlink analysis on their own in a few hours, alerting the rest of the team if necessary. Matt Wallace came up with a plan: We would still shift the uplink team schedule, but not completely around the clock. The earliest shift would start at 6:00 A.M., and the latest would end by 10:00 P.M., JPL time. The Data Controllers would still have to come in at odd hours, to look at the Mars afternoon telemetry results, but unless there was a major anomaly in the rover data, the Data Controller could go home again in a couple of hours.

With the lander team now working on Earth time, we didn't need the Rover Coordinators anymore. So we brought Art Thompson and Matt Wallace onto the uplink team as Rover Drivers/Sequence Planners. That put six engineers on the uplink team. Howard Eisen became a Data Controller, putting three engineers in that role. Eisen wrote his first Downlink Report on sol 32.

With an almost regular work schedule, and more team members to trade off responsibilities, the entire team could return to a semblance of a normal life.

Sol 35

The time had come to visit the Rock Garden. This was a group of rocks clustered together, all of them big enough to be good APXS targets. Brian Cooper designed the traverse—ambitious at almost twenty-three feet—to take Sojourner away from the lander, across a relatively clear patch of ground, then through a bumpier but still navigable stretch behind a ramp-like rock aptly called "Wedge," to halt at the Rock Garden "entrance."

The next day, Sojourner was not where we expected. Instead of driving on the far side of the rock Wedge, the rover's dead reckoning error had led it in front of Wedge. At first some of the rover team wondered how the rover could have gotten to its position: There didn't seem to be enough room between Wedge and the neighboring rocks for the rover to drive through. The drift in the rate sensor had led the vehicle astray, but its hazard avoidance behavior had been perfect. With its thread-the-needle mode enabled, Sojourner had safely navigated between Wedge and nearby Hassock rock, and found itself on the wrong side of Wedge, largely surrounded by rocky terrain.

Attempts to leave the area over the next several sols failed. The drift in the rate sensor meant Sojourner often turned to the wrong heading, and the rocks around and under the vehicle tripped safety hazard errors that aborted the traverses before the rover could get very far.

Brian Cooper unofficially dubbed the region "The Bermuda Triangle." Sojourner had gotten there with ease, but would spend days there before escaping. "We had planned to avoid this entire area, going farther to the left as we circumnavigated the lander and going directly to the Rock Garden. So here we had scientists who were very anxious for us to get to these rocks which had large vertical surfaces that were perfect for the APXS, and we were spending days and days in this other area." The frustration of the uplink team increased. "We were getting nowhere. We'd alternate. Jack would try it and I would try it, and say 'Maybe you'll get us out of here. Maybe I will.'"

The software team had built several safeguards into Sojourner, specifically to prevent the rover from putting itself at risk. These protections halted the vehicle if it encountered too steep a slope, or if the bogie angles became extreme. The design philosophy had been to allow the human operators on Earth to correct a problem before it became too severe. But once you were already in rough terrain, these same safeguards could keep you there, causing the vehicle to abort the very traverse that would lead it to safety. So, to get out of there, "we eventually tried driving to the right of Wedge . . . with the safeguards turned off." Cooper wasn't sure what would happen. To him, this was the riskiest time of the mission. "When I left for the day, I thought there was the distinct possibility—if we were unlucky—that we would find the rover flipped over the next day, and the mission would be over." Yet the mechanical team, led by Howard Eisen, insisted that the rover was capable of much more difficult terrain than the rover drivers had ever risked.

In pure Bermuda Triangle tradition, the next downlink revealed Sojourner undamaged but still at Wedge, this time with the left side wheels on top of the rock, and the right side on the ground. The uplink team chose to keep going: "We'll just drive straight off." And that's what they did. The next sol, Sojourner powered over Wedge and proceeded to the entrance of the Rock Garden.

The curse of Wedge was broken. An abundance of APXS targets were now open to us. Their names were Shark, Half Dome, Moe, and Stimpy.

Sol 56

The rover batteries died in the early morning hours of sol 56, sometime between midnight and 3 A.M. MLST. Henry Stone wrote in the End-of-Day Downlink Flight Status Report: "WE'RE ON A SOLAR MISSION FROM HERE ON OUT!" Although the batteries had survived far longer than had been promised (at sol 56 the rover had been operating for almost twice its specified "extended mission" lifetime), the loss of the batteries was still a surprise. The power engineers had been using rover telemetry to estimate battery usage throughout the mission, and their best guess had been that nearly half the batteries' capacity remained unused. Where had the rest of the energy gone? The higher the temperature, the shorter a battery's shelf life; apparently the higher than optimal temperatures during the cruise to Mars had partially discharged the batteries before the surface mission began. The power team had had no data on shelf life at the actual temperatures Sojourner had experienced, at least not until now. And since the rover had been shut down during APXS night operations, there was no hard data on power consumption for these long overnight periods, no way to make accurate quantitative measurements. They had guessed low.

How would the loss of Sojourner's batteries impact the rest of her mission? All future APXS operations would now occur during daylight hours. And the rover's alarm clock would no longer run overnight, so the rover would wake with the sun, or not at all.

By this time, the lander too was conserving its battery. While the lander's battery was being recharged each sol by its solar arrays, it was also degrading, bit by bit. So each sol, although fully charged, the lander battery stored less total energy than the sol before. To reduce power drain, the lander shut down completely each night, which meant any science data stored onboard which had not already been transmitted to Earth would be lost forever.

While the lander and rover battery problems were serious, none of us on the operations team was complaining. We were just dealing with them. These were exactly the problems we wanted to have, the ones that showed up only because everything else was working well. The rover and lander were aging, living beyond their design lifetimes. And just as with human beings, the difficulties of growing old beat the alternative.

Sojourner had made the transition to its "solar-only" mission without incident.

✳

By sol 59, we were calling it "The Thompson Loop." Sojourner sent no telemetry that day, and we thought we knew why.

Building command sequences was an involved process. The scientists' requests for mosaicked images and multiple readouts from the APXS forced sequences to be many times longer than we had predicted during Sojourner's design phase. The average sequence was now between two hundred and three hundred commands. Sometimes the Sequence Planners would have to carefully prune down the commands to fit into the part of rover memory allocated to hold sequences. Long sequences took a long time to prepare and review. By this time, the uplink team consisted of Cooper, Morrison, Welch, Wallace, Thompson, Laubach, and myself. We were all engineers who thrived on doing what we had not done before. So it was natural that we would find ways to improve the sequences, and make them easier to produce each day. Even before landing, Laubach had put together templates for the first week of operations, so the team would have something to start with, instead of building sequences from scratch. Later in the mission, we were in the habit of pulling up yesterday's sequence and modifying it for today.

Thompson's sequence for sol 58 included the first APXS experiment since the exhaustion of Sojourner's batteries. The rover was commanded to collect data from the rock dubbed Half Dome until 3:30 in the afternoon, then shut down for the night. Thompson had chosen to specify the time in Mars Local Solar Time. As usual when operating the APXS, the rover was in a communications blackout to avoid interrupting the data collection. But Thompson had tried to eke out too much data collection time: Before 3:30 P.M. came, the sun had dropped too low in the sky for Sojourner's solar array alone to provide sufficient power to sustain both the rover and its APXS. As designed, the rover began shedding power loads to protect its CPU from browning out. Finally, as its solar power waned, the rover could not maintain itself any longer, and like a computer with its plug pulled, it stopped operating. When the rover came alive in the morning sunlight of sol 59, it assumed the time was 7:15 A.M. Mars LST, the wakeup time that would have been commanded by an auto-shutdown.

Sojourner was still waiting for 3:30, only now it was early morning of the next sol. It was still in its silent mode, so there was no way to send it commands. As the operations team reasoned through what was happening, we worried that Sojourner might be stuck in its current state forever, always waiting for 3:30, always shutting off before 3:30 came . . . never completing its sequence. The Thompson Loop.

In its communications blackout state, the rover only communicated once per sol, when it woke up in the morning and asked the lander for the correct time in order to resynchronize the rover's clock. During the time synch communications session the rover wouldn't even accept new commands. If we could only tell Sojourner that it was almost 3:30, then the rover would complete the wait long before the sun went down, move on to the next command, and come out of the blackout. But we had never planned on "lying" to the rover about what time it was. The lander simply responded to a rover time request with its own estimate of the current time. The only way to send Sojourner incorrect time would be to change the lander's clock to a false value, wait for the rover to request the time, then change the lander's clock back. This would be extremely dangerous for the mission. The lander depended on knowing the time of day in order to point its high-gain antenna precisely in the direction of the Earth. Changing the lander clock would point the antenna toward the wrong point in the sky, disrupting communications. Richard Cook would not take the risk.

On sol 60 rover telemetry came down. The low-voltage condition that had existed just before the rover ceased to function on sol 58 had triggered anomalous readings of APXS current, causing the rover to declare the APXS a failed device. On sol 59, Sojourner had therefore refused to power on the APXS. Leaving the APXS off left the rover with enough power to operate until 3:30 without browning out, complete the command, and shut down normally. A sequence was uplinked for sol 60, but Thompson's original sequence had more APXS data collection built into it, so the rover was still in blackout mode, and the new sequence was not received.

Would Sojourner stay stuck in the Thompson Loop? On sol 61 the new command sequence got through, and Sojourner exited the Thompson Loop for good.

Rather than being embarrassed by his starring role in the Thompson Loop, Art would only smile. He was actually proud to have his name associated with the incident.

✳

Even driving on Mars can get tedious. Sojourner had been exploring the Ares Vallis landing site for over two months. "It was taking its toll on the operators," recalled Thompson. "Some people were starting to get fatigued and starting to almost complain about having to come in to operate this vehicle. I'd have to say I was among those people, thinking, 'How long is this going to go on?' but then I'd stop and think, 'Wait a minute. I'm driving this vehicle on Mars. Who else can get this opportunity? Will I ever get this opportunity again?' Yeah, I gotta get up, I gotta go to work, and maybe it's a fourteen-hour, fifteen-hour shift, but I'm gonna go look at a new set of pictures no one's ever seen before, and I'm gonna load the commands to make this vehicle go somewhere on another planet. That's really great!"

And as Sojourner continued its travels, it began to peer behind rocks and capture images of features that even the Pathfinder lander had never seen. Brian Cooper loved those pictures. "Some of the more spectacular images that came down were the sand dune areas beyond the reach of the IMP camera, behind the Rock Garden. That was pretty exciting. I wanted to do a lot more of that." Cooper looked forward to later in the mission, when Sojourner might venture as far as a hundred meters away from the lander. "We all wanted to see what was behind the next rock in case—you never know—there might be something really cool on the other side of that rock ... besides another rock ... some evidence ... something unknown ..."

✳

Sojourner's days began to grow shorter. It wasn't that the seasons were changing. Instead, the opportunities to communicate with the lander were beginning to shrink. For the early part of the mission, the Deep Space Network had promised Pathfinder lots of coverage from its big dish antennas situated strategically around the Earth. But Pathfinder was going on and on, and there were other customers for the DSN. Pathfinder was losing its priority. And the lander could only receive messages from the

DSN when the Earth was visible in the Martian sky. As Mars and Earth moved in their orbits, the Earth was setting earlier each sol. Sometimes, when everything else was just right, the DSN station on Earth would fail us for a maddeningly old-fashioned reason: the weather. One of the three DSN stations was located outside of Canberra, Australia. More than once, when no downlink came down from Pathfinder, the word would go out: "It's raining in Canberra." The telemetry that had streaked unimpeded across 120 million miles of interplanetary space was defeated in the last instant of its journey by a mere cloudburst.

✴

In the months before landing, some of the rover team had assumed that Howard Eisen would grab all the media attention when the time came. He often acted as if the rover was "his baby," and that no one else had nearly as crucial a role in its creation as did he. Eisen had often been the guy on the spot for interviews and rover demonstrations, both during rover development at JPL and just prior to launch down in Florida. We thought he'd maintain the pattern after landing, while the rest of the team was mostly too busy operating the rover to spend much time talking about it.

Well, this was one more misconception that did not survive the first few sols. By the time of landing, Eisen had integrated himself into the operations team. He seemed to have willingly accepted his role on the engineering analysis team as one of the key representatives of the mobility and thermal subsystem. He regularly provided insightful analysis and interpretation of the incoming data. He simply did a good job. Of course, Eisen was still Eisen. He continued to readily express strong opinions, and to tread some onto other people's turf. But if he had stopped doing that, we probably would have wondered after his health . . .

Meanwhile, the attention of the media was mostly elsewhere.

While Bill Layman had shied away from the Sojourner limelight, Donna Shirley embraced it. Just before Pathfinder landed, and then for weeks afterward, she seemed to be everywhere. She appeared on every television network news program and CNN, explaining the Pathfinder mission and how Sojourner came to be. She was widely regarded as the "mother" of Sojourner. Together with the other leaders of the Pathfinder mission—Matt Golombek, Rob Manning, Richard Cook, and newly promoted Project Manager Brian Muirhead—Shirley became a celebrity. This was an unusual role for engineers and scientists, but this group took to it, and thrived. Pathfinder had made them heroes.

Tony Spear had instead chosen to fade from view, resigning within a few days of the landing. Spear had identified two possibilities for his final act as Pathfinder Project Manager: either stand up and take the blame if Pathfinder failed, or step aside to make room for others if it succeeded. Pathfinder had succeeded. Spear appointed Brian Muirhead—up until then the deputy project manager—to take his place, and then seemingly disappeared.

The rover team had its own minor celebrity. Well before landing day, Brian Cooper had been identified in the press as *the* Rover Driver, and the rest of the rover team seemed to disappear from existence. The image of a lone engineer, video game enthusiast since childhood, choosing Sojourner's path among the treacherous rocks of Ares Vallis, was just too enticing to ignore. In much of the reporting, Cooper was represented as having complete command of the rover; even the role of the science team in selecting targets was lost. And no matter how many times he tried to explain time delay and the complexity of commanding Sojourner, most reporters could not get past their initial notion of Cooper "joysticking" the rover through the Martian desert as if it were a radio-controlled model car.

*

The engineers were growing antsy. We had been doing the bidding of the science team for over two months, and had never driven Sojourner much more than a dozen yards away from the lander. As even Matt Golombek acknowledged, "We'd been in the Rock Garden. We'd spent weeks going inches each day, and it was very frustrating. They were sick of going to rocks, and they wanted to go drive ten meters." When were we going to get the chance to go somewhere? The rover team pressed the scientists to allocate some time to long rover traverses. Golombek described the incident as the "Outburst of the Rover Drivers." Golombek laughed, characterizing the rover team's attitude: "'Come on, let's just put the pedal to the metal.'" Matt Wallace led the "insurrection."

The result was a new long-range rover plan. After a few more sols in the Rock Garden, the rover team would take Sojourner for a spin. We would perform several long traverses, and from these we would learn Sojourner's true capabilities. This would give us the chance to exercise the rover's mobility and navigation performance over longer distances than we had ever had time for during our Earth-bound testing of Marie Curie.

※

On sol 84 we didn't hear from the lander.

In recent sols, the lander had been commanded to shut down completely overnight. Its battery, although rechargeable unlike Sojourner's, was known to be near the end of its useful life. That was okay: The lander could operate in a solar-power-only mode just as Sojourner was already doing, at least for a while. The lander team had estimated that they could get one more night of useful science out of the batteries. For the night of sol 83, the lander was instructed to turn itself back on after midnight. When Pathfinder was silent on the morning of sol 84, the team speculated that the lander's battery had gone dead sometime during the Martian night. When the battery failed, so had the lander's electronic clock. The CPU lost track of time. The lander depended on its clock to tell it what time of day it was and, therefore, where in the sky to point its high-gain antenna to reach the Earth. Most likely, the high-gain antenna was pointing in the wrong direction. But the lander had a separate low-gain antenna, which operated at a much lower data rate and did not need to be carefully pointed. It should be possible to reestablish contact with Pathfinder through that low-gain antenna.

The days went by and still we did not hear from the lander.

※

Sojourner had contingency sequences onboard, designed to kick in if the normal sequence running on the rover terminated and no new sequence arrived from Earth. I had written the original contingency sequences a year before, about three months before launch. They had been loaded on-board before Sojourner was shipped to the Cape.

From a rover-centric point of view, the worst of all possible worlds would be one in which the Pathfinder lander made a perfect landing, opened its petals, and the rover was dead on arrival. The media reaction would have been "All that work to get to Mars, all that money, and the rover is just going to sit there forever!" As the sols went by, with the IMP camera operating flawlessly, more and more pictures would come down, documenting the rover's continuing inactivity in full color and at higher and higher resolution.

If the failure had been communications-induced, the original contingency sequences would have avoided that nightmare by causing the rover to begin a fully autonomous mission a few days after landing. The sequences would have commanded the rover to circumnavigate the lander at a range of about fifteen feet, blindly searching for APXS soil and rock sites, attempting to image the lander, and transmitting the data back in the hope that someone was still listening.

As the mission progressed, the original contingency plan became irrelevant, and we started to worry that a combination of minor glitches might activate the contingency sequence when all we wanted was for the rover to sit still. So, on sol 80, we uplinked a new contingency sequence to replace the old one. Now if we lost touch with Sojourner, it would sit where it was for six sols, then drive back toward the lander. Since the lander, ramp, and airbags were a hazard to the rover, we had created a keep-out zone around the lander. The rover would try to reach a way-point located at the center of the lander, while the keep-out zone kept it from getting too close. The combination created an unreachable target. The rover would drive in a circle around the lander, at a range of about ten feet, again and again. This would go on until the rover received a new command sequence, or encountered a hazard it couldn't handle.

Four sols later, we were out of touch with the Pathfinder lander.

TWENTY-FIVE

WILL BUILD SPACECRAFT
FOR FOOD

We never regained contact with the lander or Sojourner. The Pathfinder team attempted to reestablish communications daily for weeks, then tried again weekly, and at last monthly. After the final failed attempt, the project threw a party—a Pathfinder wake. Even with the loss of contact on sol 84, Pathfinder had been a success beyond any of our expectations.

The mission had largely disappeared from the media's attention before it ended. Yet Pathfinder had entered the fabric of the national consciousness. During the three months of surface operations, the Pathfinder web site received nearly 750 million hits; some people claimed that this unprecedented web traffic represented the Internet coming of age. The old lament "If we can send men to the Moon, why can't we . . ." now had an updated version: "If we can put a skateboard on Mars . . ." Sojourner showed up in political cartoons, on the comics pages, and even in an episode of an animated television series. And the U.S. Postal Service issued a stamp commemorating the mission, barely five months after landing day.

Although we were disappointed to see the mission end, the Sojourner team was already chomping at the bit to move on to something new. We were designers, developers, and implementers, not maintenance engineers. We were not meant to do the same thing every day, even operating a robotic vehicle on the surface of Mars. By the time the Pathfinder lander spoke to us for the last time, we were already planning to train new people to take our places, so we could begin work on the next mission.

The 2001 rover mission was to be more ambitious than Pathfinder by far. The rover would be wholly a different beast. It would be bigger, faster, and far more complex. Where Sojourner stayed within about fifty feet of its lander and drove barely a hundred yards, the rover for 2001 was intended to move up to five *miles* across the Martian terrain. It would dispense with support from the lander altogether, communicating directly with an orbiter that would swing by overhead for a few minutes twice each sol. Rather than merely analyze rocks in place on the surface, the new rover would drill core samples and retrieve them for eventual return to Earth. And instead of promising a week-long mission, we were committed to operating on Mars for a full Earth year.

Most of the Sojourner team moved on to the new rover project. And as we began work on this next effort, we found even greater challenges than with Sojourner. JPL had been ordered by NASA to reduce its workforce by over one thousand people in the two years to come. The in-house capabilities to machine and produce hardware continued to dwindle in the face of this mandate. The new rover team was asked to create a more capable vehicle in less time than MFEX, with about the same funds. At the start of Pathfinder, the rover had been just a payload, and some people on the project had even hoped to eliminate it completely. The common wisdom—both at NASA and JPL—had been that the "faster, better, cheaper" Pathfinder mission could not be done; yet the Pathfinder team had always believed in itself. With the success of Pathfinder, the institutional viewpoint had transformed again: The new rover and its instruments were now the centerpiece of the mission. It seemed that everyone in JPL management and at NASA headquarters believed the ambitious 2001 mission could be accomplished. Only those of us charged with carrying it out were unsure.

The Pathfinder and Sojourner team members had become victims of their own success. What had been the exception became the rule. Future missions were now expected to cost even less and do still more. Pathfinder was held up as the shining proof that this expectation was realizable. Other projects at JPL talked about "Pathfinderizing" themselves, but no one who had been on Pathfinder ever used that term. The keys to Pathfinder's success—largely the quality of its people, their freedom to cut across usual organizational boundaries to solve problems, and the mandate to modify the scope of the mission as necessary to remain within its fixed budget—seemed to be lacking in these other efforts. At a panel discussion on "How Pathfinder Invented Faster, Better, Cheaper" held in Washington, D.C., on November 5, 1998, the Project Manager was asked, "What would you do differently if you had the chance to do Pathfinder again?" Brian Muirhead smiled and answered. "That's not the question. The trick is being able to do the same thing again." The environment at JPL had changed.

In conversations with the individual Sojourner team members after the mission was over, I encountered a recurring theme, an almost universal recognition of how unique the rover team had been.

For Rick Welch, who had joined the team after launch, his one regret was not having been involved from the beginning. "To me, the cradle-to-grave approach just seems fun, first of all. I mean, building it . . . Which is more fun, making the taco or eating it? I sort of like doing both. I think having the opportunity to actually operate your own rover creation, I would think would be very satisfying. Why would you want to build something that you actually couldn't carry right through and see how it operates, and see the success of it?" And despite the future uncertainties, JPL was still the place to be. ". . . You can't compare JPL to anyplace else. It is the only place that builds robotic spacecraft. It is the only place, now, that builds rovers. For the moment. What else would you want to do?" Welch paused. "It's still hard to believe that we're actually on Mars."

*

A year after Landing Day, the next rover mission was in turmoil. The requirements had been changed and changed again. Given the combination of limited funds, too little time, and ambitious science goals which we had not been allowed to compromise, the consensus of the rover team was that we could not deliver what had already been promised by others. For months, the Mars Exploration Directorate would not accept that message. Or, if it did, the engineers in the trenches were not getting any feedback to tell them that the problem was being dealt with. The new rover team did not have the momentum that had been so critical to the Sojourner team. We weren't a locomotive barreling down the track; it was as if our train kept returning to the station.

It turned out the Mars Exploration Directorate had heard at least some of the rover team's concerns. The Directorate knew it was in trouble. As originally envisioned, the 2001 mission included the rover with its science payload, the lander that put the rover down on the surface and had its own independent set of science instruments, and a science / communications relay orbiter circling the planet. Even if each of these elements was made to cost less than Pathfinder, there just wasn't enough money allocated to the Mars Surveyor Program to put all of these spacecraft together by the 2001 launch date. JPL managers went to NASA headquarters for relief, hoping to eliminate or put off some of the 2001 objectives to make the mission manageable in scope. Headquarters sent them packing. JPL had pulled off Pathfinder, hadn't they? They had successfully built a rover before; why should the next rover require so much more money? Why the sudden attack of conservatism? Headquarters just didn't believe that JPL couldn't do the job they'd signed up for.

An independent review board was convened to determine what the Directorate could and should be doing. Former Pathfinder Project Manager Tony Spear, who was no longer working on Mars missions, led the board. The board's conclusion: Indeed, the 2001 mission had "too much on its plate" and could not hope to complete an orbiter, lander, and rover with full capabilities for the time and money available. They recommended delaying the 2001 rover to the next launch opportunity, in 2003. And they further advised that the Mars Surveyor Program should focus much more of its resources on its stated objective of returning Martian rock and soil samples to the Earth.

So, after six months of planning and development of the 2001 rover, the entire structure of the Mars Surveyor Program was up for grabs. Study teams formed to determine how best to redesign the program and keep it within budget. Workshops were set up to give industrial firms the opportunity to suggest alternatives. There was even talk and political pressure to keep a rover on the 2001 mission by refurbishing and flying the Sojourner test rover, Marie Curie.

To me, amid these uncertainties, the single-mindedness of Pathfinder seemed to have been lost. By the first anniversary of the day Sojourner and the Pathfinder lander had successfully arrived at Ares Vallis, how could so much have changed?

❊

At almost exactly the same time, a NASA ceremony was held at JPL, honoring those responsible for the major achievements of the previous year. Individuals and groups associated with a variety of JPL activities and projects were being acknowledged, but one of the Pathfinder engineers sitting next to me said, "It looks like a Pathfinder reunion!" As I looked around, I saw that it was true. I hadn't seen so much of the team together in one place in months.

Chairs had been set up in rows on the mall, under a tent that shaded guests and honorees from the summer sun. We faced toward the steps of the JPL administration building. On the first landing had been set a podium, and behind it tables stacked with awards to be handed out. From large speakers came familiar music from the soundtracks of several science fiction movies. And everybody was dressed up!

The music ended. Dr. Ed Stone, the JPL Director, and Dr. Wes Huntress, the NASA Associate Administrator for Space Science, presided. Huntress told us how much healthier space science was at NASA, compared to early in the decade. Today, JPL had "more missions on the books than ever" in its history. Then he started listing those missions, and the crowd cheered as each was named. Finally it was time for the honorees to be presented with their awards. As each recipient was named and the specifics of the award were read aloud, the individual would have the medal draped around his or her neck, and would then have a photograph taken flanked by Dr. Stone and Dr. Huntress.

Among the other recipients, each of the members of the Sojourner core team was awarded the NASA Exceptional Achievement Medal. But two awards were the most pleasing to me. For his invention of the rocker-bogie, Don Bickler was the sole recipient of the rare Exceptional Engineering Achievement Medal. And Hank Moore, wearing a suit jacket, tie—and his Sojourner baseball cap—was presented with the Exceptional Service Medal for his participation in the Pathfinder mission. This was nearly the identical award he had been given twenty-one years earlier for his work on the Viking mission to Mars.

After the ceremony, we all wandered around congratulating each other. I even found myself congratulating Tom Economou, the terror of the APXS. Matt Golombek turned to look at the two of us and started to laugh. "Look at you guys! You're actually smiling at each other. I don't think I've ever seen this before!" I glanced from Matt to Tom to Matt, and gave Tom a slap on the back. "That's because this is the first time you've seen us not working together!"

There were buses waiting to take us to the Caltech President's Residence for a luncheon. We all looked a bit silly wearing our medals, but we didn't care. When we arrived at the luncheon, the President of Caltech shook hands with each of us. We found another tent set up on the lawn, in which were tables with white tablecloths. Carafes of iced tea and lemonade had been placed on each table. And there was a buffet of Vietnamese spring rolls, salmon, sesame asparagus, chicken, and fresh melon.

The people sitting at my table had all shared the experience of Pathfinder and Sojourner: Allen Sirota, Art Thompson, Tom Economou, Justin Maki, Bill Dias. Some had with them their spouses or children, who had also experienced the mission, but in a different way. Sitting next to me was Sharon Laubach, who since Pathfinder had become my fiancée.

We were just engineers who had done our jobs well. But here we were, sharing a luncheon with the President of the California Institute of Technology and the Director of JPL. We were wearing medals as if we were Olympians. And as someone at the table commented, here was the free lunch. All I could think was that, for someone working at JPL and sending probes into deep space, this was as good as it gets.

The group at the table talked about the mission. Tom Economou wondered whether Sojourner might be operating even now, still collecting APXS spectra and sending them back to the mute lander.

The mission seemed a very long time ago.

Justin Maki raised his glass for a toast. Glasses of lemonade and iced tea went up around the table. "Here's to doing it again."

And that, of course, would be the biggest reward of all.

INTERLUDE

The 2001 big rover mission was rescheduled to 2003. Less grandiose missions—a lander and an orbiter—were begun to fill the open launch window in 2001. Marie Curie was added as a payload to the 2001 Mars lander mission. Meanwhile, the 2003 rover mission was recast as Mars Sample Return, intended to bring Mars rocks back to Earth.

Then the two "faster, better, cheaper" missions to Mars launched in 1998—another lander-orbiter pair—failed. The combined cost of the two missions had been less than the single Pathfinder mission. The NASA Administrator admitted that "faster, better, cheaper" had gone too far.

In the reassessment that followed, the 2001 Mars lander mission was cancelled, and Marie Curie went back into a box. Mars Sample Return was also eliminated. It appeared that no mission to Mars would launch during the 2003 opportunity.

Then a few Pathfinder alumni proposed a new mission to rise from the ashes of the others. By combining the Pathfinder lander design with the original 2001 big rover, it might still be possible to deliver a long-range rover to Mars by early 2004. NASA agreed, and the Mars Exploration Rover mission was born. NASA headquarters asked one more thing of JPL: Build not one, but two identical rover spacecraft, to be sent to two distinct landing sites and operated simultaneously.

On February 1, 2003, the Space Shuttle Columbia disintegrated during reentry over Texas, killing the seven astronauts onboard. Despite new uncertainties over the future of the space program in the tragedy's aftermath, two rovers, direct descendants of Sojourner and Pathfinder, are on their way to Mars.

MARS REDUX

The sting of two Mars mission failures only months before was still fresh as the Mars Exploration Rover mission (MER) got underway in May of 2000. JPL and NASA had learned at least one lesson from the cases of "faster-better-cheaper" taken to the extreme. This time, the project would be given the financial resources to do the job right.

Yet, MER would face other challenges. On Pathfinder, the overarching constraint—the 800-pound gorilla that influenced every decision—had been budget; for MER, it would become time. The alignment of the planets would force MER to launch in the summer of 2003, so according to our new schedule, we'd be roving on Mars barely three and a half years after the new mission started. That gave us significantly less development time than Pathfinder . . .

How were we going to make this work?

The MER mission had been sold—and its feasibility argued—on the premise that the lander would be largely "build to print" from Pathfinder. This meant that most of the new spacecraft would be unchanged from the Mars Pathfinder design. The notion was appealing: by adapting a flight-proven spacecraft design and a partially designed rover, MER would have a head start, with much of the design work already complete. There should be plenty of time to complete the design and build the spacecraft before the upcoming launch date, with margin to spare.

But the devil was in the details. As the design effort proceeded, the mass and volume of the rover that could accommodate the accepted science instrument payload began to increase. The new spacecraft was going to be heavier than Pathfinder; this fact would ripple through the entire design, forcing the use of new materials and creative solutions. The MER rover would weigh 384 pounds to Sojourner's twenty-two, and would have ground clearance enough to drive right over Sojourner, perhaps just scratching the smaller rover's solar array. The mechanical team struggled with how to fit the rover into the cramped lander, which would require folding it up like an intricate origami creation, far more complex than Sojourner's deployment.

New estimates for the velocity of the winds on Mars now led us to believe that during its landing, Pathfinder may have been lucky—the chance of high winds was greater than we had once imagined, and these winds could have smashed the lander into the rocks hard enough to rip the cushioning airbags and damage the spacecraft. MER would therefore need a new sensing subsystem to detect such winds, and rockets to counteract them in the seconds before dropping onto the Martian surface. Only then would the likelihood of a safe landing be great enough to be worth the risk.

In response to the growing engineering challenges, the development schedule was replanned, monitored, and adjusted. The intensity of the project effort rapidly escalated. Costs increased, more personnel joined the team, and the number of hours worked per person per day rose—all to keep us on schedule for the launch dates that could not move.

❋

MER was designed to perform impressive science investigations. Two identical rovers, eventually named "Spirit" and "Opportunity," would not collect samples as the cancelled Mars Sample Return would have done; instead, they would study the rocks on-site, using as many science instruments as they could carry with them. The mission objective was to "follow the water"—to determine from the geology of Mars whether liquid water had ever existed in abundance on the surface, providing one of the key conditions conducive to life.

The new rover concept effectively melded together the capabilities of Sojourner and the Pathfinder lander. It was almost as if Sojourner had been asked to carry the lander on its back as it roamed the surface of Mars. The rover itself would contain the radio equipment to communicate directly to Earth, as well as to the two active spacecraft now orbiting Mars, which would act as relays. The rover would carry its own mast-mounted stereo camera for science, but with three times the resolution of Pathfinder's IMP. Additional cameras on the front and rear of the rover body and on the mast would enable rover navigation and hazard detection. Each rover was expected to survive for a minimum of three months while driving to distinct geological sites. At those sites, the rover would deploy a robotic arm with about the reach of a human arm for placing instruments on rocks and soil, including an APXS like Sojourner's for assessing the elemental makeup of science targets and another sensor for determining the mineralogy of iron-bearing rocks. Perhaps most appealing to human eyes, the robotic arm carried a microscopic imager, often called a "geologist's hand lens" that could be placed within an inch or less of rock surfaces for high-resolution close-up photos. The instrument arm could also deploy a "rock abrasion tool" to grind through the outer layer of rocks, exposing the potentially unweathered material beneath for perusal by the other arm-mounted instruments.

While Sojourner had been merely a payload on the Pathfinder lander—a small hitchhiker along for the ride—the MER rover would occupy the entire internal volume of the new lander: The brain of the rover would direct all spacecraft functions on the way to Mars. When the rover rolled off the lander, it would leave behind a dead hulk, a mere platform on which the rover had rested before beginning its explorations.

※

Henry Stone joined MER as the manager of electronics and software development—similar to his role on Sojourner, but now scaled up to encompass the entire spacecraft destined for Mars. I became manager of the operations system—the combination of software, teams, and processes that would operate the MER spacecraft during cruise and then direct the rover after landing. Other ex-Sojourner team members—Art Thompson, Allen Sirota, Jake Matijevic, Matt Wallace, Sharon Laubach, and Rick Welch among them—signed on to MER in various capacities. Many Pathfinder personnel took key management positions.

MER rapidly evolved into a juggernaut. While the members of the Sojourner team had worked harder during that mission than at any other time in their careers, the new project demanded more—a sustained effort of fifty-, sixty-, even eighty-hour work weeks for almost three years. And the tone of the Mars Exploration Rover mission was strikingly different from Pathfinder: where Pathfinder had been entrepreneurial, MER was more . . . corporate. Yet the large group of engineers that comprised the MER team was just as dedicated as the Pathfinder and Sojourner teams, and stepped up to the project's needs. The engineers on the new team really wanted to get to Mars—and for most of them it would be for the first time.

MER did several things to make up for lost time. The engineering team was bigger—more than three times that of the less ambitious Pathfinder. At MER's peak, more than a thousand people were on the payroll. A new culture evolved within the new team. The project seemed to be holding meetings constantly. These meetings were a necessary evil, engendering within the team a shared understanding of the mission objectives and evolving design. Engineers took to carrying laptop computers and cell phones everywhere, to continue their design work even during the many meetings. It seemed the only way to keep up with the daunting pace of the project. Wireless networks were being set up throughout the JPL facility, providing connectivity in every conference room on-Lab. The MER team took full advantage of this new capability. People paid attention in meetings only when the most relevant topics were being discussed. Otherwise, they were lost in the details of their own work, captured on their laptops. In meetings with thirty people in the room, it became commonplace to ask engineers the same technical question twice: once to get their attention, then again to get an answer.

✻

Spirit and Opportunity had been given specific science objectives and traverse requirements, which would be challenging to meet in their presumed ninety sol lifetimes. As the operations development manager, I was charged with defining how we would operate these two new rovers. At a design review, I presented an approach in which the operations team would again live on Mars time, sending commands to the rovers every sol and responding rapidly to what had happened to keep the mission moving steadily forward to achieving its objectives. Only this time, we would build a bigger team, with enough people to operate the two rovers simultaneously seven days a week while giving everybody regular days off. The review board first told me that the fact that we had commanded Sojourner every sol was no proof that it could be done on a "real" mission, and then turned around and admonished me that any changes to the way we had operated Pathfinder would have to be carefully justified.

The operations approach was not the only aspect of MER that came under close scrutiny. Throughout the project, review boards without fail told MER management that its hardware and software development and integration schedule was impossibly ambitious. Finally, at the last review before Spirit was to reach Mars, one of the board members admitted his personal embarrassment over telling the project for three years that what we were doing couldn't be done, and our constant success in proving him wrong.

Of course, those of us on the MER team hadn't been sure the schedule was possible either.

✻

The two launches were scheduled about two and a half weeks apart, in early and late June. Spirit would go first. There had always been concerns over a dual launch: the two spacecraft were on adjacent pads, and the second one to go would have to wait for the first to be on its way before final preparations could be made. If unanticipated problems significantly delayed Spirit, Opportunity's chance could be jeopardized. But there was no real drama here: Spirit lifted off on the third day of its launch period, as soon as the weather at the Cape cooperated.

Drama and frustration came from another source. There was nothing wrong with the Opportunity spacecraft itself. But the launch vehicle—a new variant of the Delta that had sent Pathfinder to Mars—had a series of minor if nagging problems: a battery went dead and had to be replaced, an area of cork insulation wouldn't stay glued down, a fuel valve was sluggish. The combination of these occurrences delayed Opportunity for eleven days, until there were only seven days left in the launch period. On July 7, another glitch scrubbed the launch with eight seconds to go. The launch team scrambled to recycle the countdown for the second attempt of the night, exactly forty-two minutes and fifty-two seconds later. This time, the launch was perfect. Few MER team members witnessed it, however, as most of the people who had flown out to see the launch had long since returned to JPL. (My wife and I were driving across country, and watched the launch on a laptop computer in a motel room in the middle of the New Mexico desert.)

After the second launch, the cruise of the two spacecraft to Mars was almost eerily routine. While part of the team focused on getting the twin spacecraft to Mars, others worked the choreography of how we would unfold the rover and safely drive it off the lander, which experience had already shown would take over a week. A third group learned how to operate the rovers once they had bid the landers good-bye—which this time would require about twenty people working side by side over each night to build the commands for all of the instrument observations and rover activities that would take place the next Martian morning.

The spacecraft were being aimed at two landing sites halfway around the planet from each other, both considered good candidates for revealing evidence of past water on Mars. Spirit was headed for a January 3 landing at Gusev crater. From orbit, Gusev looked like an ancient crater lake nearly a hundred miles wide, with a huge channel emptying into the crater. When it arrived three weeks later, Opportunity would land in Meridiani Planum, a broad flat expanse that showed signs of the presence of the mineral gray hematite, which on Earth usually formed in the presence of water.

⁂

After the failure of the last attempted Mars landing in 1999, NASA public relations had worked hard to prepare the press and the public for the possibility of another failure this time. The press kits pointed out how many things could go wrong. The JPL Director described the entry, descent, and landing of MER as "six minutes of terror." And even if the landing were successful, we might have to wait until the Spirit rover's second sol on Mars to hear from it for the first time.

But, thanks to the radio tones the spacecraft transmitted every time a major onboard event occurred, we were able to follow the progress of Spirit all the way down to the surface. Then, immediately after the signals indicated the lander had hit the ground, contact was lost. Even though this was expected—the lander would be bouncing and rolling on Mars for a third of a mile or more before coming to rest, making communications difficult—everyone tensed up. Would we hear from Spirit again?

Ten minutes later, we got another signal. Spirit had survived! And even landed upright, with its antenna pointed at the sky.

Amid all the cheering I was still worried. Several things still had to go right: the retraction of the airbags, opening of the lander petals, and then the critical step of getting the solar arrays deployed. Without them, Spirit would be starved for energy, and would soon die. With them, we would have the time we needed to solve any problems. Earth had set in the sky over Gusev crater, so direct communication with Spirit was no longer possible. The soonest we could know about the critical deployments would be when Odyssey—one of the science spacecraft in orbit over Mars we would use as a communications relay—passed over the landing site and sent back to Earth any data received.

The Odyssey data was coming in. People in the control area cheered this news: If any data were being downlinked, it meant that Spirit had succeeded in communicating with the orbiter on its first attempt—a very good sign. Then the engineering telemetry came in . . . the first picture— a shot of a calibration target . . . then more pictures!

None of us could believe our luck. The rover looked perfect, with its solar panels fully extended, and the camera mast fully deployed. All the engineering data looked nominal. There were no fault conditions—much better than any of our rehearsals!

In only a minute or two, the ground software had constructed a panorama from the individual images that had come down. We could see 360 degrees around the rover, to the horizon. The terrain was flat with lots of small rocks. My reaction was almost exactly the same as seven years earlier when Pathfinder revealed its own landing site: "We can drive here!"

※

At an all-hands meeting a few days before Spirit's landing, the JPL Director asked if there was anything he could do to show his appreciation for all the team had already accomplished in getting the mission this far. One of the mission managers spoke up: "Free ice cream!"

A few sols after the landing, a commercial freezer with a sliding glass top appeared outside the break room, filled to the top with boxes and boxes of ice cream bars.

JPL management was soon amazed at how much ice cream the operations team was consuming per week. I knew that some engineers were so busy working the ten hour Mars-time shifts that they were on the "MER diet" of two or three ice cream bars a day. And a few of them were still losing weight, since this was all that they were eating . . .

Every time the freezer would go low, it was restocked. We wondered how long the ice cream would last. We joked that if Opportunity's landing attempt failed, the freezer would be gone the next day.

※

"Six wheels on dirt!"

With airbags in the way of a straight drive off, unsuccessful attempts to retract them out of the way, and the general caution of the operations team, it took us until sol 12 to finally get Spirit's wheels dirty.

Spirit then immediately began its post-egress exploration of Mars, analyzing the soil immediately in front of the rover. Everything was working great. We did out first drive to a rock the scientists had named "Adirondack." We were going to analyze the rock with our instruments, and then grind away a coin-shaped depression in the rock's surface with the rover's rock abrasion tool. When this was done, we'd apply our robotic arm-mounted instruments to the hole.

During the night of sol 17, the uplink team built the commands to grind our first hole in a Martian rock.

✳

On sol 18, we couldn't get most of our command sequences to the rover. At first we thought that the thunderstorms at our transmitter in Australia were getting in the way of our commands. But as the day wore on, our ability to get commands to the rover and see the right responses seemed to deteriorate. Sometimes the rover didn't answer when expected. The rover would miss communications sessions, and the team would try to develop an explanation. Then we'd get a report that there was a lot of data coming down from Spirit through one of our relay orbiters; the team would cheer with relief, only to discover minutes later that the content of the data was meaningless. What was going on? There had been no warning of a problem onboard. And our mission had only just gotten started!

When you can't communicate with your spacecraft, engineers worry. A lot. There's no way to tell whether the problem can be easily fixed, of if this is the beginning of the end for the mission. And when the mission did end—whether it was a day or a year from now—it would be when we asked the rover to talk to us and got back only silence. The newspaper headlines soon read: "Spirit rover in critical condition!"

Over the next sol or two we commanded Spirit to send us "beeps" that would prove it was listening to us. Sometimes these worked.

After one such attempt, we got no signal. The mood in the control room collapsed into desolation. The team forced itself into planning what to do next.

A few minutes later, there was a somewhat tentative, incredulous voice on the network: "Uh. Flight. Telecom. Station 63 is reporting carrier lock." Engineers around the room looked up in surprise. "They're reporting symbol lock . . . We've got telemetry." Spirit was back! The data coming down was garbled, but the rover was at least babbling at us. The mood in the room had transformed again.

From what little information we could get, it seemed that Spirit was booting up and resetting over and over, never fully waking up, never completely shutting down. When the rover did come up long enough to begin communicating, it might reset in the middle of talking to us. And whatever was causing the resets was preventing Spirit from doing the rest of its tasks, like preparing its data to be sent. The power and thermal engineers were particularly worried: Spirit was designed to shut down and cool off each night. In its current state, it might never be turning off. The result might be that the rover was overheating and draining its batteries down to nothing. On sol 20, an attempt to command the rover to shut itself down for the night failed—Spirit might be listening to us, but wasn't often doing what it was told.

*

I came back on shift as the Tactical Uplink Lead on the night of sol 20. The team decided that the objective for sol 21 would be to make Spirit shut down. The software team had been meeting all day, coming up with scenarios to explain the rover's bizarre behavior. Their best notion was that the rover's main "flash" memory was corrupted, causing the rover to reset every time it tried to access that memory. So, the software team proposed forcing the rover to wake up without using the flash memory, like booting a computer without using its hard drive.

On the morning of sol 21 we sent the appropriate commands. The rover started talking to us on schedule. We were getting data that made sense again! The telemetry confirmed that the batteries were nearly depleted. We sent a command to tell the rover to go to sleep until the next morning. This would give Spirit time to recharge its batteries. All the indications were that Spirit had listened to us this time. We upgraded the rover's condition to guarded, but stable.

We still had a lot of work to do. But now that Spirit was communicating, we could begin to trace the problem. With a few sols we proved that the rover's malady was triggered when there were too many data files in its flash memory. All we had to do was keep a close eye on the number of onboard files, and we could prevent the problem from ever recurring.

By sol 33 Spirit was fully restored and we were exploring Gusev crater again.

❋

In the meantime, Opportunity had been falling toward Mars. On the night of Saturday, January 24, those of us working on Spirit paused long enough to watch the landing events unfold. Waiting for a Mars landing is always nerve-wracking. (After Pathfinder and Spirit, this was my third time.) But this was as flawless as it could be. Data came back from Opportunity pretty much the whole time, during descent, landing, and bouncing across the Martian surface! And a few hours later, the Odyssey orbiter relayed photos of our new landing site that were amazing, even for Mars. There were no rocks! We had rolled to a stop at the bottom of a small crater. The soil was a grayish red, except where the lander airbags had disturbed it; in those spots it looked like a deep, pure red. There were detailed imprints (even revealing seams) of the airbags in the soil in several places. And while there were no individual rocks, Opportunity seemed to be partly encircled by a rock outcropping—bedrock. No one had ever seen this on Mars before. And it was only yards away. One of the lead scientists standing next to me in the mission control area seemed in awe. "Jackpot!"

Opportunity had landed on its side. The airbags affixed to the main and forward lander petals were then hanging in the air, instead of trapped underneath the lander as had happened for both Pathfinder and Spirit. The result was that the airbags had been fully retracted before Opportunity had righted itself: there was no airbag material at all in the way of driving forward off the lander. It appeared that we had a perfectly clear path off the lander into an unexpectedly alien landscape.

With this stroke of luck and the operations team's recent Spirit experience, Opportunity drove off the lander earlier than planned—on sol 7. The rover was destined to spend two months exploring the rock outcrop inside its tiny crater.

. . . and the ice cream continued.

❋

We all knew that a key press conference was coming—important enough that it would take place at NASA headquarters rather than at JPL. The results to be announced were so sensitive—at least from a scientist's perspective—that if you missed the one meeting at which it was discussed with the operations team, the science team wouldn't breathe a word of it. It turned out that NASA wanted to be certain of its conclusions—as verified by an independent panel of scientists—before any rumors began circulating.

On March 23, 2004, scientists associated with the Mars Exploration Rover mission announced that at least part of the Meridiani Planum landing site had once been under water. A mosaic of photos of part of the outcrop taken by Opportunity's microscopic imager had provided telltale signs of flowing saltwater at least several inches deep, in which the rock had formed. The presence of liquid water implied that Mars was likely once much warmer, probably with a much more substantial atmosphere—otherwise the water would have frozen or quickly evaporated.

Ancient Mars was indeed beginning to sound more and more like a habitable environment.

The NASA Associate Administrator declared that these results put the Meridiani site "at the top of the list" for where to land the future Mars Science Laboratory mission, which would have instruments for life detection. "These are the kinds of rocks that are exceptionally good at preserving evidence of microbial life." So if you wanted to find microscopic fossils, this would be the place to look. The assessment was tantalizing: The Opportunity rover might be sitting in the middle of fossil evidence of past life on Mars, but did not have the instruments onboard to detect the fossils if they were there.

Reporters at the press conference wanted to know when we would definitively answer the question of whether life had ever existed on Mars. Would it be days, weeks, a few months? The assembled scientists smiled at the question, for they knew that MER would not settle it. These rovers simply weren't designed for it. At best, the answer might come after several more years.

And for that, we would have to go to Mars . . . again.

GLOSSARY

ACS	Attitude Control System
ADM	APXS Deployment Mechanism
AI	Artificial Intelligence
APXS	Alpha Proton X-ray Spectrometer
ATLO	Assembly, Test, and Launch Operations
Caltech	California Institute of Technology
CARD	Computer-Aided Remote Driving
CCD	Charge-Coupled Device; solid-state camera
CMU	Carnegie-Mellon University
CPU	Central Processing Unit
CRT	Cathode Ray Tube; picture tube
DARPA	Defense Advance Research Projects Agency
DDF	Director's Discretionary Fund
DSN	Deep Space Network
EDL	Entry, Descent, and Landing

EEPROM	Electrically Eraseable Permanent Read-Only Memory
EOWG	Experiment Operations Working Group
ETL	U.S. Army Engineer Topographic Laboratories
FET	Field Effect Transistor
FUR	Flight Unit Rover; also known as Sojourner
g	1 Earth gravity
GALCIT	Guggenheim Aeronautical Laboratory, California Institute of Technology
GM	General Motors
GSE	Ground Support Equipment
HGA	High-Gain Antenna
IMP	Imager for Mars Pathfinder
JATO	Jet-Assisted Take-Off rocket
JPL	Jet Propulsion Laboratory
KSC	Kennedy Space Center
MESUR	Mars Environmental SURvey
MFEX	Microrover Flight Experiment
MLST	Mars Local Solar Time
MRSR	Mars Rover Sample Return
MSA	Mission Support Area; Mission Control for Pathfinder operations
MSM	Mars Science Microrover
NASA	National Aeronautics and Space Administration
NATO	North Atlantic Treaty Orgranization
ORT	Operations Readiness Test
RCW	Rover Control Workstation
RHU	Radioisotope Heater Unit
RORT	Rover Operations Readiness Test

RTTV	Robotic Technology Test Vehicle
SAN	Semi-Autonomous Navigation
SDM	Software Development Model rover
SGI	Silicon Graphics Incorporated
SIM	System Integration Model; also known as the Marie Curie rover
SLIM	Surface Lander Investigation of Mars
SLRV	Surveyor Lunar Roving Vehicle
TACOM	U.S. Army Tank Automotive Command
USGS	United States Geological Survey
VOCA	Voice Operated Communications Assembly
WEB	Warm Electronics Box

DRAMATIS PERSONAE

Don Bickler	JPL mechanical engineer; inventor of the rocker-bogie; member of Sojourner operations team
Gary Bolotin	Rover lead electronics engineer
Brian Cooper	Software architect and developer of the Rover Control Workstation; Rover Driver during Mars operations
Howard Eisen	Leader of the rover mechanical, thermal, and mobility team; member of the rover downlink team during Mars operations
Matt Golombek	Mars Pathfinder Project Scientist
Ken Jewett	Rover mechanical engineer; responsible for Sojourner's overall configuration
Arthur "Lonne" Lane	Task manager of the Mars Science Microrover demonstration
Sharon Laubach	Caltech graduate student; member of Sojourner operations team
Bill Layman	Rover Chief Engineer
Jake Matijevic	Rover lead system engineer; promoted to Sojourner team leader

David Miller	JPL robotics engineer; early proponent of the micro-rover concept
Andrew Mishkin	Rover system engineer; Sojourner Sequence Planner
Jack Morrison	Lead software architect and developer for Sojourner's onboard software; Rover Driver during Mars operations
Tam Nguyen	Sojourner software engineer
Jim Parkyn	JPL communications engineer
Glenn Reeves	Lander lead software engineer
Carl Ruoff	Supervisor of the JPL Robotics group
Donna Shirley	Sojourner team leader; promoted to director of the Mars Exploration Program
Allen Sirota	Rover system engineer; integration and test lead engineer; Data Controller during Mars operations
Tony Spear	Pathfinder Project Manager
Henry Stone	Leader of the rover control and navigation team; Data Controller during Mars operations
Scot Stride	Rover communications engineer
Lin Sukamto (Lin van Nieuwstadt)	Leader of the Sojourner telecommunications subsystem team during rover development
Jan Tarsala	Rover communications engineer during Mars operations
Art Thompson	Rover system engineer; Rover Coordinator during operations
Matt Wallace	Rover power subsystem engineer; Rover Coordinator during operations
Rick Welch	Sojourner Sequence Planner and Rover Driver
Al Wen	Lead Sojourner thermal engineer
Bob Wilson	System engineer for the Mars Science Microrover demonstration
Brian Wilcox	Supervisor of the JPL Robotic Vehicles group; inventor of CARD; member of Sojourner operations team

ACKNOWLEDGMENTS

When I started work on this book, I imagined it to be a solitary effort. By now it has become clear just how many people have contributed to its success. Significant accomplishment is rarely achieved by one individual acting alone. As with the development of Sojourner itself, the deeper one looks, the more contributors one finds.

I am indebted to the many rover team members who provided their recollections or reviewed portions of the text for accuracy, including Brian Wilcox, Henry Stone, Matt Wallace, Allen Sirota, Brian Cooper, Art Thompson, Jack Morrison, Ken Jewett, Scot Stride, Gary Bolotin, Rick Welch, Jake Matijevic, Bill Dias, Sharon Laubach, Hank Moore, Joy Crisp, Don Bickler, Howard Eisen, Jan Tarsala, and Bill Layman. Other participants in the Pathfinder mission or the rover-related history herein also allowed themselves to be subjected to interviews: Bob Anderson, Justin Maki, David Gruel, Matthew Golombek, Arthur L. Lane, Robert Wilson, and Ken Manatt.

Friends, associates, and relatives read drafts and gave me important feedback: Bill Hicks, Jan Ludwinski, Deborah Bass, Mark Adler, Andy Morrison, and Bobbie Laubach. Mary Forgione provided special editorial assistance in reducing the originally voluminous manuscript to a more manageable length.

Several people at JPL helped in the process of tracking down photographs, dealing with permission issues, and making the photos available. Sue LaVoie, Xaviant Ford, Grace Fisher-Adams, Michael Jameson, Jeanne Rademacher, Tom Thaller, and David Deats each provided invaluable assistance.

I offer thanks to my agent, Agnes Birnbaum, for her persistence, advice, and faith in the book; and to my editor Natalee Rosenstein at The Berkley Publishing Group. Esther Strauss at Berkley always responded cheerfully to my seemingly unending stream of questions regarding various details of production.

I want particularly to thank my wife, Sharon, for her patience during the long hours of writing and editing, when I often disappeared completely into another world. I've come back!

INDEX

Page numbers in italics indicate illustrations; those in bold indicate tables.

PHOTOGRAPHIC CREDITS

All photographs in this book, unless otherwise noted, are courtesy NASA/Jet Propulsion Laboratory/California Institute of Technology. Images of the Sojourner rover, the names Sojourner ®, Mars Rover ®, and the spacecraft design are copyright © 1996–97, California Institute of Technology, with all rights reserved, and further reproduction prohibited.

The unofficial Sojourner rover patch design is courtesy Calvin Patton.

The diagram of the solar system was created by the author. The positions of the planets on July 4, 1997, were determined using SOAP (Satellite Orbit Analysis Program) software developed by the Aerospace Corporation.